Thermal Analysis with SOLIDWORKS Simulation 2018 and Flow Simulation 2018

Paul M. Kurowski, Ph.D., P.Eng.

CERTIFIED
Solution
Partner

3S SOLIDWORKS

SDC
Publications

Design Generator, Inc.

SDC Publications
P.O. Box 1334
Mission, KS 66222
913-262-2664
www.SDCpublications.com
Publisher: Stephen Schroff

About the Cover
The image on the cover shows results of thermal analysis of HEAT SINK TH model introduced in chapter 10.

ISBN-13: 978-1-63057-165-8
ISBN-10: 1-63057-165-2

Printed and bound in the United States of America.

About the Author

Dr. Paul Kurowski obtained his M.Sc. and Ph.D. in Applied Mechanics from Warsaw Technical University. He completed postdoctoral work at Kyoto University. Dr. Kurowski is an Assistant Professor in the Department of Mechanical and Materials Engineering, at the University of Western Ontario. His teaching includes Finite Element Analysis, Product Design, Kinematics and Dynamics of Machines and Mechanical Vibrations. His interests focus on Computer Aided Engineering methods used as tools of product design.

Dr. Kurowski is also the President of Design Generator Inc., a consulting firm with expertise in Product Development, Design Analysis, and training in Computer Aided Engineering.

Dr. Kurowski has published many technical papers and taught professional development seminars for the Society of Automotive Engineers (SAE), the American Society of Mechanical Engineers (ASME), the Association of Professional Engineers of Ontario (PEO), the Parametric Technology Corporation (PTC), Rand Worldwide, SOLIDWORKS Corporation and others.

Dr. Kurowski is a member of the Association of Professional Engineers of Ontario and the Society of Automotive Engineers. He may be contacted at www.designgenerator.com.

Acknowledgements

I thank my students for their valuable comments and questions.

I thank my wife Elżbieta for project management and support in writing this book.

Paul Kurowski

Table of contents

Before You Start

Notes on hands-on exercises and functionality of Simulation

This book goes beyond a standard software manual. It takes a unique approach by bridging the theory of heat transfer with examples showing the practical implementation of thermal analysis. This book builds on material covered in **"Engineering Analysis with SOLIDWORKS Simulation."**

We recommend that you study the exercises in the order presented in the book. As you go through the exercises, you will notice that explanations and steps described in detail in earlier exercises are not repeated in later chapters. Each subsequent exercise assumes familiarity with software functions discussed in previous exercises and builds on the skills, experience, and understanding gained from previously presented problems. Exceptions to this sequential approach are chapters 14, 15, 16 where **Flow Simulation** is used. These chapters may be skipped by readers not interested in problems involving **Flow Simulation**.

Exercises in this book require different **SOLIDWORKS Simulation** functionality and the functionality depends on which **SOLIDWORKS Simulation** product is used. The **SOLIDWORKS Simulation** Product Matrix document is available at:

http://www.SOLIDWORKS.com/sw/products/simulation/simulation-matrices.htm

All exercises in this book use **SOLIDWORKS** models which can be downloaded from www.SDCpublications.com. Most exercises do not contain any **Simulation** studies; readers will be asked to create studies, results plots, and graphs themselves. All problems presented here have been solved with **SOLIDWORKS Simulation Premium** running on Windows 7 64 bit.

You may find your results slightly different from results presented in this book. This is because numerical results may differ slightly depending on the operating system and software service pack.

We encourage you to explore each exercise beyond its description by investigating other options, other menu choices, and other ways to present results. You will soon discover that the same simple logic applies to all functions in **SOLIDWORKS Simulation**, be it structural or thermal analysis.

This book is not intended to replace software manuals. Therefore, not all Thermal Analysis capabilities with **SOLIDWORKS Simulation** are covered. The knowledge acquired by the reader will not be strictly software specifics. The same concepts, tools and methods apply to any FEA program.

Prerequisites

"**Thermal Analysis with SOLIDWORKS Simulation**" is not an introductory text. Rather, it picks up Thermal Analysis from where it was left in the pre-requisite textbook "**Engineering Analysis with SOLIDWORKS Simulation**." If you are new to **SOLIDWORKS Simulation** we recommend using "**Engineering Analysis with SOLIDWORKS Simulation**" to gain essential familiarity with Finite Element Analysis and **SOLIDWORKS Simulation**. At the very least go through chapters 1, 2, 3, 7, 8, 10 of that pre-requisite textbook.

The following prerequisites are recommended:

- ❏ An understanding of Heat Transfer Analysis
- ❏ An understanding of Structural Analysis
- ❏ Familiarity with **SOLIDWORKS** CAD
- ❏ Familiarity with **SOLIDWORKS Simulation** to the extent covered in "**Engineering Analysis with SOLIDWORKS Simulation**" or equivalent experience
- ❏ Familiarity with the Windows Operating System

Selected terminology

The mouse pointer plays a very important role in executing various commands and providing user feedback. The mouse pointer is used to execute commands, select geometry, and invoke pop-up menus. We use Windows terminology when referring to mouse-pointer actions.

Item	Description
Click	Self-explanatory
Double-click	Self-explanatory
Click-inside	Click the left mouse button. Wait a second, and then click the left mouse button inside the pop-up menu or text box. Use this technique to modify the names of folders and icons in **SOLIDWORKS Simulation** Manager.
Drag and drop	Use the mouse to point to an object. Press and hold the left mouse button down. Move the mouse pointer to a new location. Release the left mouse button.
Right-click	Click the right mouse button. A pop-up menu is displayed. Use the left mouse button to select a desired menu command.

All **SOLIDWORKS** file names appear in CAPITAL letters, even though the actual file names may use a combination of capital and small letters. Selected menu items and **SOLIDWORKS Simulation** commands appear in **bold**; **SOLIDWORKS** configurations, **SOLIDWORKS Simulation** folders, icon names and study names appear in *italics* except in captions and comments to illustrations. **SOLIDWORKS** and **Simulation** also appear in bold font. Bold font may also be used to draw reader's attention to a particular term.

1: Introduction

Topics covered

- ❑ Heat transfer by conduction
- ❑ Heat transfer by convection
- ❑ Heat transfer by radiation
- ❑ Thermal boundary conditions
- ❑ Analogies between thermal and structural analysis
- ❑ Thermal elements: solids and shells
- ❑ Scalar and vector entities, presenting results
- ❑ Steady state thermal analysis
- ❑ Transient thermal analysis
- ❑ Linear thermal analysis
- ❑ Nonlinear thermal analysis

What is Thermal Analysis?

Thermal analysis deals with heat transfer in solid bodies. We approach thermal analysis from the perspective of a user experienced in structural analysis such as static, modal, buckling etc. as implemented in **SOLIDWORKS Simulation**. You will soon notice that experience in structural analysis is directly transferable to thermal analysis because of the close analogies between structural and thermal analyses. The temperature is analogous to displacement in structural analysis, strain to temperature gradient, and stress to heat flux. Selected analogies are summarized in Figure 1-1.

Structural Analysis	Thermal Analysis
Displacement [m]	Temperature [K]
Strain [1]	Temperature gradient [K/m]
Stress [N/m^2]	Heat flux [W/m^2]
Load [N] [N/m] [N/m^2] [N/m^3]	Heat source [W] [W/m] [W/m^2] [W/m^3]
Prescribed displacement [m]	Prescribed temperature [K]
Pressure [N/m^2]	Prescribed heat flux [W/m^2]
Hook's law: $$\sigma = E\frac{du}{dx}$$	Fourier's law: $$q = -k\frac{dT}{dx}$$
Stiffness matrix	Conductivity matrix

Figure 1-1: Analogies between structural and thermal analyses with units in SI system.

The primary unknown in structural analysis is displacement; the primary unknown in thermal analysis is temperature. This leads to an important difference between structural and thermal analysis performed with the finite element method. Displacement, which is a vector and includes both translation and rotation, requires up to six degrees of freedom per node. The number of degrees of freedom in structural analysis depends on the type of elements; for example, solid elements have three degrees of freedom per node and shell elements have six degrees of freedom per node. Two-dimensional structural elements have two degrees of freedom per node. Temperature is a scalar and requires only one degree of freedom per node, regardless of element type. This makes thermal problems much easier to solve because thermal models typically have fewer degrees of freedom as compared to structural models.

Another conceptual difference is that thermal analysis is never a "static" analysis. If heat flow does not change, then the problem is "steady state analysis" and not static because heat flow never stops. If heat flow changes with time, then the problem is called transient.

Mechanisms of heat transfer

Conduction

In a solid body, the energy is transferred from a high temperature region to a low temperature region. The rate of heat transfer per unit area is proportional to the material thermal conductivity, cross sectional area and temperature gradient in the normal direction; it is inversely proportional to the distance (Figure 1-2). This mode of heat transfer is referred to as conduction:

$$Q_{COND} = kA(T_{HOT} - T_{COLD})/L$$

Where:

Q_{COND} – Heat transferred by conduction [W]

k – Thermal conductivity [W/m/K]

A – Cross sectional area [m^2]

T_{HOT} – Temperature on the hot side [K]

T_{COLD} – Temperature on the cold side [K]

L – Distance of heat travel [m]

Conduction is responsible for heat transfer inside a solid body.

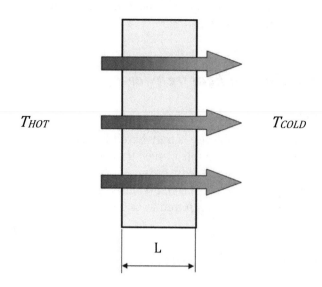

Figure 1-2: Heat transfer by conduction.

Conduction is responsible for heat transfer inside a solid body.

Thermal conductivity is vastly different for different materials as shown in Figure 1-3.

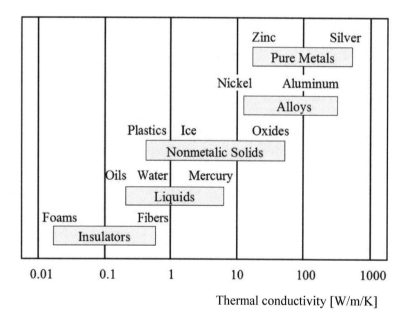

Figure 1-3: Thermal conductivity for different materials.

There are five orders of magnitude difference in thermal conductivity between the best conductors and best insulators.

Convection

Convective heat transfer is the heat flow between a solid body and the surrounding fluid (either liquid or gas). Convective heat transfer can be either natural convection where the fluid flow is due to the variation in specific weight of a hot and cold fluid, or forced convection where the fluid is forced to flow past the solid body. Therefore, natural convection requires gravity (Figure 1-4), and forced convection does not require gravity (Figure 1-5). Since fluid (air, water, steam, oil etc.) is required for heat transfer by convection, this type of heat transfer cannot happen in a vacuum. Heat exchanged by convection is expressed as:

$$Q_{CONV} = hA(T_S - T_F)$$

Where:

Q_{CONV} – Heat transferred by convection [W]

h – Convection coefficient [W/m^2/K]

A – Surface area [m^2]

T_S – Surface temperature [K]

T_F – Fluid bulk (ambient) temperature [K]

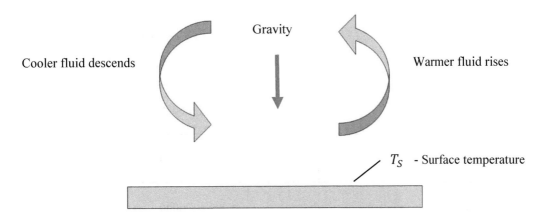

T_F – Bulk temperature of fluid in contact with surface

Gravity

Cooler fluid descends   Warmer fluid rises

T_S – Surface temperature

Figure 1-4: Heat transfer by natural convection.

Convective heat transfer can take place only in the presence of fluid and gravity.

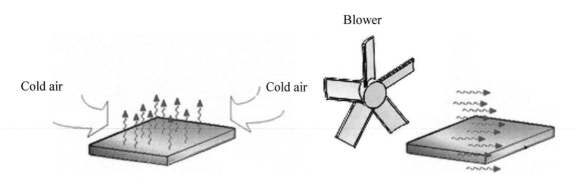

Figure 1-5: Natural and forced convection.

Forced convection does not require gravity.

The magnitude of the convection coefficient strongly depends on the medium (fluid) surrounding a solid body (Figure 1-6).

Medium	Convection coefficient $W/m^2/K$
Air (natural convection)	5-25
Air/superheated steam	20-300
Oil (forced convection)	60-1800
Water (forced convection)	300-6000
Water (boiling)	3000-60000
Steam (condensing)	60000-120000

Figure 1-6: Convection coefficient for different media.

There are five orders of difference in convection coefficients between different media and different types of convection.

Radiation

Radiation heat transfer occurs between a solid body and the ambient or between two solid bodies without presence of any medium (fluid). This is the only type of heat transfer that occurs in a vacuum. Heat flows by electromagnetic radiation. The upper limit to the emissive power is a back body radiating heat and is prescribed by the Stefan-Boltzmann law:

$$q = \sigma T_s^4$$

Where:

q – Heat flux (heat emitted by radiation per unit of area) [W/m^2]

σ – Stefan-Boltzmann constant = 5.67x10^{-8} [W/m^2/K^4]

T_s – Surface temperature [K]

The heat flux emitted by a real surface is less than that of a black body at the same temperature. It is given as:

$$q = \varepsilon \sigma T_s^4$$

where ε is a radiative property of a surface called emissivity. Values of ε are in the range of $0 \leq \varepsilon \leq 1$ and provide a measure of how well the surface emits radiative energy in comparison to a black body. It depends strongly on the surface material and finish. For example, a polished aluminum surface has an emissivity of about 0.05; an oxidized aluminum surface has an emissivity of 0.25. The emissivity also depends on the temperature of the face emitting heat by radiation.

In heat transfer by radiation, heat may be radiated out to space by a single body; it may be exchanged between two bodies or it may be exchanged between two bodies as well as radiated out to space (Figure 1-7).

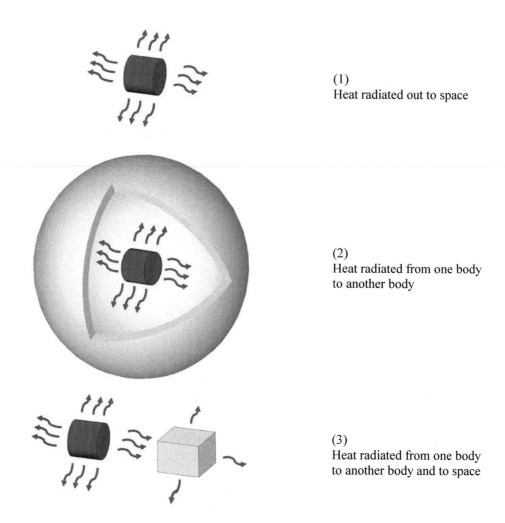

(1)
Heat radiated out to space

(2)
Heat radiated from one body
to another body

(3)
Heat radiated from one body
to another body and to space

Figure 1-7: Different cases of heat transfer by radiation.

If heat is exchanged between two bodies only and not radiated out to space, then one body must be fully enclosed by the other body (2). Heat gained by radiation may be radiated out again into space (3).

Having introduced three mechanisms of heat transfer we need to make a very important statement. With **SOLIDWORKS Simulation** which uses the finite element method, only heat transfer by conduction is modeled directly. Convection and radiation are modeled as boundary conditions. This is done by defining convection and/or radiation coefficients to faces that participate in heat exchange between the model and the environment.

The next two examples show how a prescribed temperature, convection and heat load work together to induce heat flow inside a solid body. These examples are not hands-on exercises. Hands-on exercises will start in chapter 2.

Heat Flow Induced By Prescribed Temperatures

Just like stresses may be caused by prescribed displacement, heat flow may be induced by temperature differences defined by prescribed temperatures. Consider the model BRACKET TH with different temperatures defined on two faces as shown in Figure 1-8. Note that the temperature field establishes itself in the model but heat flow continues due to temperature gradients. Also notice that no heat escapes from the model because we have not defined any mechanism to exchange heat through any surfaces other than the two faces with prescribed temperatures. This implies that the model is perfectly insulated, except for two faces where prescribed temperatures are defined and may correspond to a situation where the bracket holding a hot pipe is mounted on a cold surface and heat escaping through faces exposed to air is negligible.

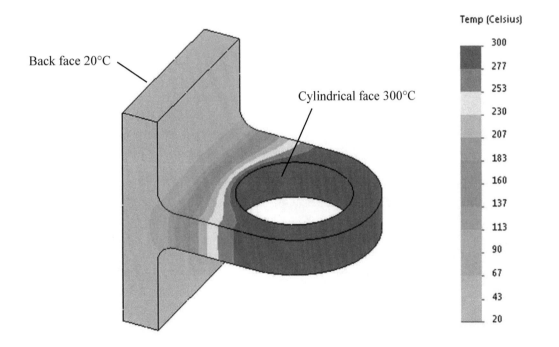

Figure 1-8: Temperature distribution in the model where heat flow is induced by prescribed temperatures.

A temperature of 300°C is applied to the cylindrical face, and a temperature of 20°C is applied to the back face.

Plot in Figure 1-8 uses custom colors (grey substituted for blue) to improve the black and white print quality. Custom colors will be used frequently to present fringe plots in this book. Custom colors may be defined in **Chart Options** in plot settings.

Heat flow induced by heat load and convection

Heat flow can also be induced by the applied heat load. The unit of heat load applied to a surface is called heat flux; heat flowing through an imaginary cross-section is also called heat flux. Total heat applied to a volume or face is called heat power. Notice that since thermal analysis deals with heat flow, a mechanism for that heat flow to occur must be in place. In the heat sink problem HEAT SINK01, shown in Figure 1-9, heat enters the radiator model through the base, as defined by the applied heat power which is a close analogy to force load in structural analysis. Convection coefficients [W/m^2/K], also called film coefficients, are defined for all remaining surfaces and provide the way to remove heat from the model; ambient temperature must be also defined.

Heat enters the model through the base where heat power is defined. Heat escapes the model through faces where convection boundary conditions are defined; definition of convective boundary conditions consists of definition of convection coefficients and ambient temperature, which is temperature of the surrounding fluid.

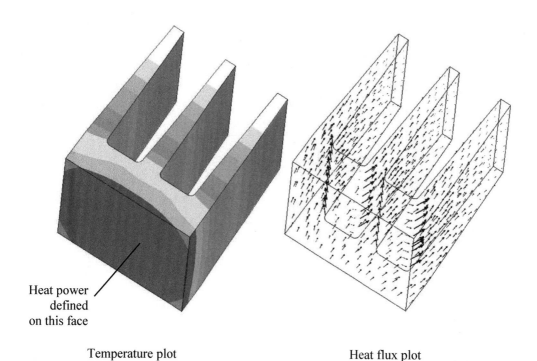

Heat power
defined
on this face

Temperature plot   Heat flux plot

Figure 1-9: Temperature distribution and heat flux in a heat sink model.

A temperature plot being a scalar quantity can only be shown using a fringe display. Heat flux is a vector quantity and can be illustrated either by a fringe plot or by a vector plot. Notice that arrows "coming out" of the walls illustrate heat that escapes the model because of convection.

The structural analogy of convection coefficients is a bit less intuitive. Convection coefficients are analogous to elastic support offered by distributed springs. Just like supports and/or prescribed displacements are necessary to establish model equilibrium in a structural analysis problem, convection coefficients and/or prescribed temperatures are necessary to establish heat flow in a thermal analysis problem. Indeed, an attempt to run thermal analysis with heat loads but with no convection coefficients or prescribed temperatures results in an error similar to the one caused by the absence of supports in structural analysis.

Modeling considerations in thermal analysis

Symmetry boundary conditions can be used in thermal analysis based on the observation that if symmetry exists in both geometry and boundary conditions, then there is no heat flowing through a plane of symmetry. After simplifying the model to ½ in case of single symmetry or to ¼ in case of double symmetry (Figure 1-10), nothing needs to be done to surfaces exposed by cuts. No convection coefficients defined for those surfaces means that no heat flows across them.

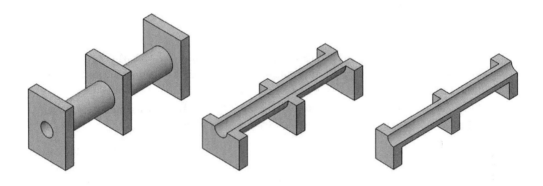

Full model   ½ model   ¼ model

Figure 1-10: Model DOUBLESYM with double symmetry can be analyzed using ¼ of the model.

No thermal boundary conditions are applied to faces in the plane of symmetry.

Axisymmetric problems may be represented by 2D models as shown in Figure 1-11.

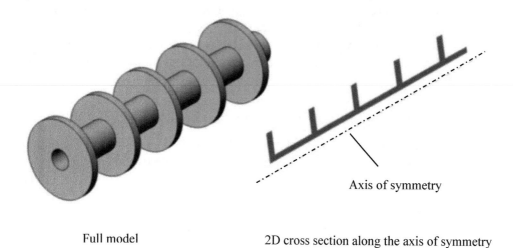

Axis of symmetry

Full model   2D cross section along the axis of symmetry

Figure 1-11: Axisymmetric model AXISYM represented by a 2D cross section.

Notice that models DOUBLE SYM (Figure 1-10) and AXISYM (Figure 1-11) are suitable for analysis of temperature, but because of sharp re-entrant edges, they are not suitable for analysis of heat flux near the sharp re-entrant edges because heat flux there is singular. This is in direct analogy to sharp re-entrant edges causing stress singularities in structural models.

Once you acquire sufficient familiarity with thermal analysis you are encouraged to analyze models BRACKET TH, HEAT SINK01, DOUBLE SYM, AXISYM using thermal parameters of your choice.

Let's wrap up this review with an important observation which will serve as a guide to all thermal analysis problems. In structural analysis we define a load path by defining loads and supports. In a heat transfer problem, we define a mechanism of heat transfer; we must know how heat enters the model, how it travels through the model and how it exits the model.

Types of thermal analysis

Thermal analysis may be linear or nonlinear, steady state or time-dependent as shown in Figure 1-12.

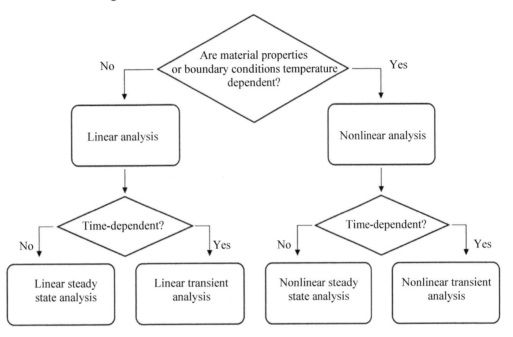

Figure 1-12: Types of thermal analyses.

In linear analysis, the conductivity matrix does not change during the solution process; in nonlinear analysis it must be modified because material properties and/or boundary conditions are temperature dependent. In steady state analysis, material conditions and boundary conditions do not change with time; in time dependent (transient) they do change.

Thermal analysis with SOLIDWORKS Simulation

SOLIDWORKS Simulation uses methods of finite element analysis to solve both structural and thermal problems. CAD models prepared in **SOLIDWORKS** are discretized (meshed) into finite elements which type depends on the type of geometry prepared in **SOLIDWORKS**. Solid bodies are meshed into solid elements; Surface bodies are meshed into shell elements; 2D axisymmetric and extruded models are meshed into corresponding 2D elements. Types of finite elements, meshing techniques and other mesh specific considerations are discussed in detail in the pre-requisite text "**Engineering Analysis with SOLIDWORKS Simulation**." Here we limit our review to thermal analysis specific issues.

Elements available in thermal analysis with **SOLIDWORKS Simulation** are shown in Figure 1-13. Notice that beam elements are not available in thermal analysis.

	3D elements		2D elements
	Solid elements	Shell elements	Axisymmetric elements Extruded elements
First order element Linear (first order) temperature field Constant heat flux field			
Second order element Parabolic (second order) temperature field Linear heat flux field			

Figure 1-13: Elements available in thermal analysis with SOLIDWORKS Simulation.

The majority of analyses use the second order solid tetrahedral elements.

First order elements do not offer any advantages in either structural or thermal analysis and beam elements are neither applicable nor available in thermal analysis. Therefore, we are left with second order solids and second order shells to model 3D problems and second order plate elements to model 2D axisymmetric or extruded problems.

Prior to commencing work with thermal problems, make sure that **Solid Bodies** and **Surface Bodies** are visible in the **SOLIDWORKS** Feature Manager (Figure 1-14).

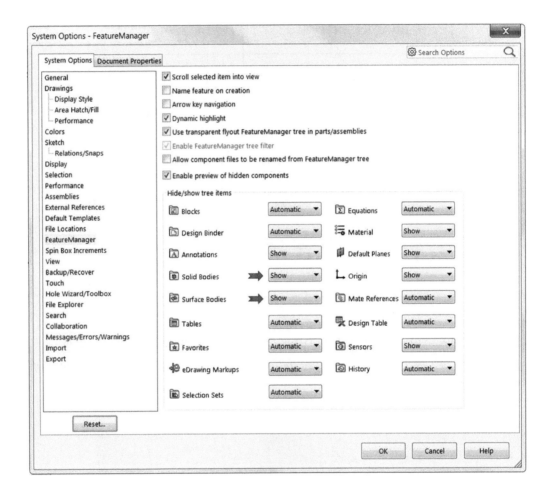

Figure 1-14: Settings of Feature Manager.

To facilitate working with Simulation models make Solid Bodies and Surface Bodies always visible.

Models used in this chapter

Model	Configuration
BRACKET TH	*Default*
HEAT SINK 01	*Default*
DOUBLE SYM	*01 full*
	02 half
	03 quarter
AXISYM	*Default*

Figure 1-15: Names of models and model configurations used in this chapter.

You are encouraged to analyze these models using thermal analysis settings of your choice.

2: Hollow plate

Topics covered

- ❑ Heat transfer by conduction
- ❑ Heat transfer by convection
- ❑ Different ways of presenting results of thermal analysis
- ❑ Convergence analysis in thermal problems
- ❑ Solid elements in heat transfer problems
- ❑ Shell elements in heat transfer problems

Project description

We will conduct thermal analysis of a simple model to study the effects of discretization error and the use of different types of elements. We start with HOLLOW PLATE TH, similar to the model from the introductory textbook **"Engineering Analysis with SOLIDWORKS Simulation"** where it is used to demonstrate structural analysis.

Open model HOLLOW PLATE TH and review the two configurations: *01 solid* where the model is represented as a solid body and *02 shell* where the model is represented as a surface body. Stay in the *01 solid* configuration and create a thermal study called *01 solid*. Apply the prescribed temperature boundary conditions as shown in Figure 2-1; these prescribed temperatures will induce heat flow from hot to cold.

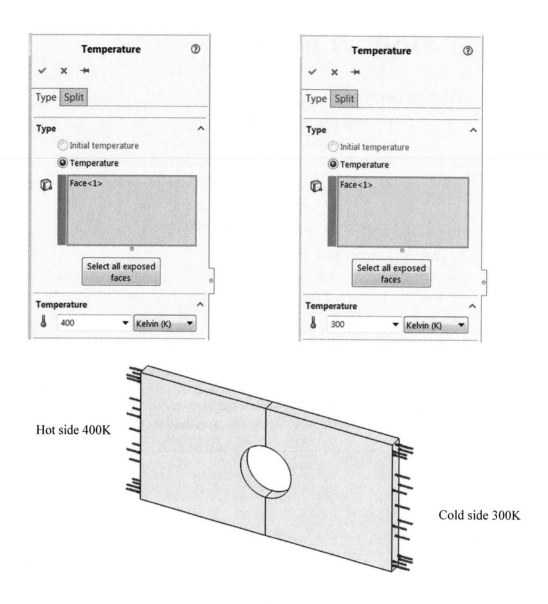

Hot side 400K

Cold side 300K

Figure 2-1: Prescribed temperatures boundary conditions applied to two end faces.

No convection is defined anywhere in the model meaning that all faces are insulated except where temperature boundary conditions are defined. Split faces are added to the model to facilitate sensor location. Later, they will be used to define convection to one side only.

Repeat the definition of the prescribed temperature on the hot side where it is 400K. Mesh the model with a coarse mesh as shown in Figure 2-2.

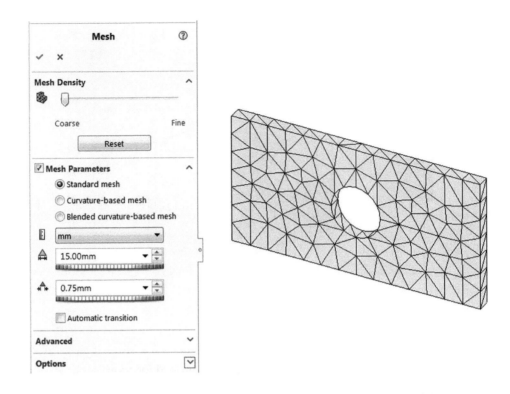

Figure 2-2: Coarse mesh created with 15mm element size.

Use Standard mesh; do not use Draft quality elements. This mesh is coarse; it is used only as the first step in the convergence process.

Run the solution and create two plots: **Temperature** and **Resultant Heat Flux**, and probe results as shown in Figure 2-3 and Figure 2-4.

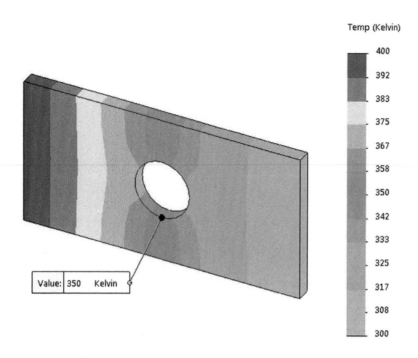

Figure 2-3: Temperature results: 350K in the probed location.

Probe at the vertex created by the split line.

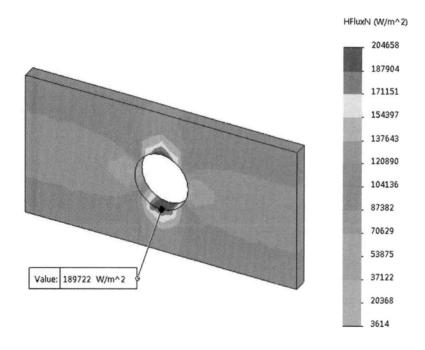

Figure 2-4: Resultant heat flux results: 189722W/m² in the probed location.

Probe at the vertex created by the split line. Your result may be slightly different depending on the solver and service pack used. This applies to all Simulation results presented in this book.

Repeat the analysis using a mesh of default size 5.72mm (study *02 solid*) and a fine mesh with element size 2mm (study *03 solid*). Optionally try meshing the model with 1mm element size but be prepared for a long run (study *04 solid*). Notice that while temperature results are almost insensitive to mesh refinement, the maximum resultant heat flux changes with mesh refinement as shown in Figure 2-5.

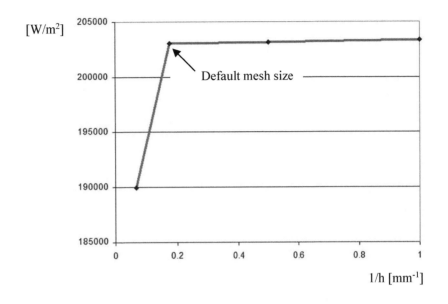

Figure 2-5: Resultant heat flux in the probed location as a function of the inverse of the element size.

Four points are connected by straight lines to visually enhance the graph.

As the graph in Figure 2-5 demonstrates, the heat flux converges to a finite value and the default element size produces acceptable results. What should also be noticed is that every given mesh introduces artificial thermal resistance; the coarser the mesh, the larger that added resistance is, and that causes lower heat flux. This is the effect of discretization error in direct analogy to structural models where artificial stiffness is added to the model by discretization (meshing).

Copy study *02 solid* (the one with the default mesh) into study *05 solid convection*. We will use it to demonstrate convective heat transfer out of the plate. Delete prescribed temperature on the cold side. In its place, define **Convection** as shown in Figure 2-6.

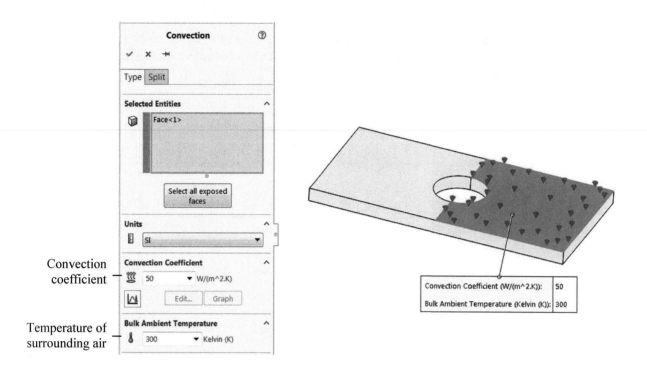

Figure 2-6: Definition of convection on ½ of one side.

As always, definition of convection requires a convection coefficient and a bulk temperature, which is the temperature of the fluid surrounding the model.

A convection coefficient of 50W/m² /K corresponds to free convection with air.

Run the solution of study *05 solid convection* and review results shown in Figure 2-7.

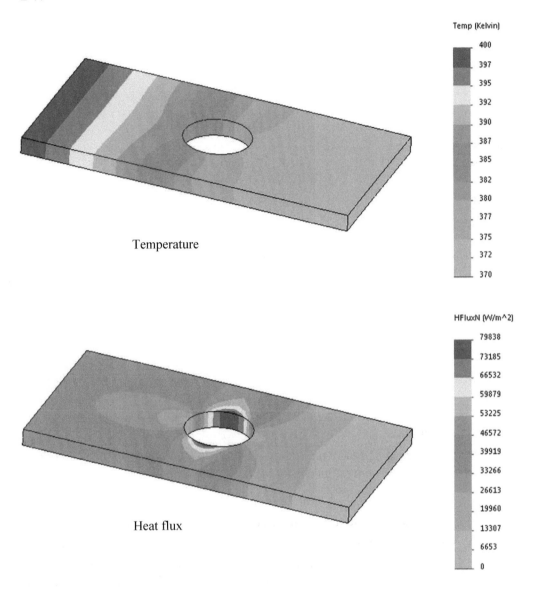

Figure 2-7: Temperature and heat flux results in the presence of convection.

Temperature is a scalar value; it can be plotted only using a fringe plot as shown above. Heat flux is a vector value; presenting it as a fringe plot shows only the magnitude but not the direction of heat flow.

Modify the heat flux plot into a vector plot as shown in Figure 2-8.

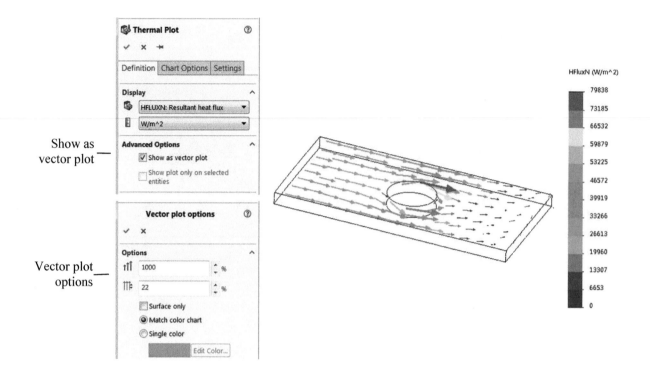

Show as vector plot

Vector plot options

<u>Figure 2-8: Heat flux results presented as a vector plot display the direction of heat flow.</u>

Rotate the plot as required to see vectors "coming out" of the model through the face where convection conditions have been defined.

You may want to review the results of other studies with heat flux results shown as a vector plot.

Having demonstrated analogies between the convergence process in structural and thermal analyses using solid elements, we will now study the use of shell elements in thermal analysis.

Stay with model HOLLOW PLATE TH but switch to *02 surface* configuration. Create a thermal study *06 surface convection*. Exclude the Solid Body from analysis and define the surface thickness as shown in Figure 2-9.

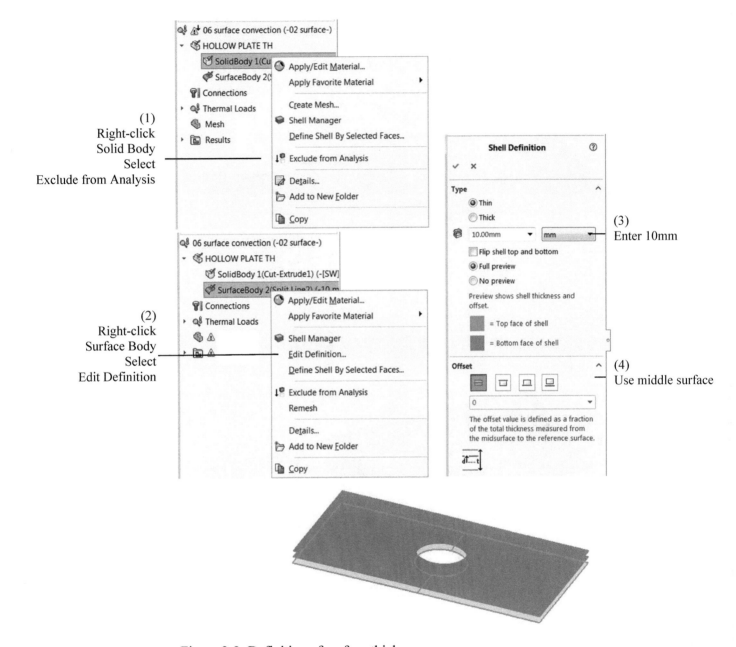

Figure 2-9: Definition of surface thickness.

Even though the Solid Body is invisible in the CAD model, it still forms a part of the CAD geometry and needs to be excluded from the analysis.

The definition of thickness is required because thickness is not specified in the surface CAD model. Thermal analysis does not distinguish between Thin and Thick shell element formulation.

Apply thermal boundary conditions as in study *05 solid convection*: a prescribed temperature of 400K on the hot side (the edge) and convection to the side face as shown in Figure 2-6. Use the default element size to mesh the model into shell elements, obtain a solution and present **Temperature** and **Heat Flux** results using fringe plots (Figure 2-10).

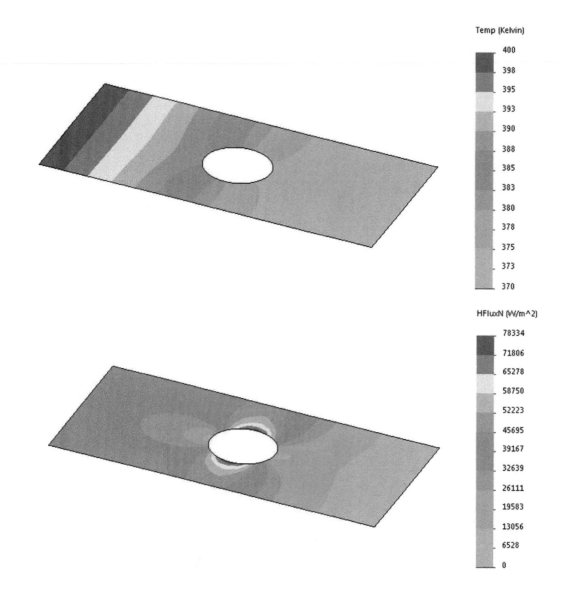

Figure 2-10: Temperature and heat flux results from the shell element model presented as fringe plots.

These results are very similar to the results produced by the solid element model.

While fringe plots show little difference between results obtained from the solid element and shell element models, vector plots do show important differences (Figure 2-11).

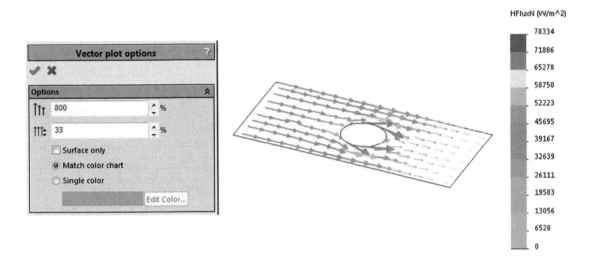

Figure 2-11: Heat flux results in the shell element model presented using vector plot.

Rotate the plot to see that heat flux vectors do not "come out" of the model.

In structural analysis, shell elements differentiate between top and bottom and different stress results are read on opposite sides. In thermal analysis, shell elements do not differentiate between sides. Therefore, the shell element model cannot show to which side heat escapes from the model. As shown in Figure 2-11, heat flux vectors diminish without "coming out" of the model.

Since in thermal analysis shell elements do not differentiate between sides, you cannot apply different thermal boundary conditions to two sides of a face meshed with shell elements.

Summary of studies completed

Model	Configuration	Study Name	Study Type
HOLLOW PLATE TH.sldprt	01 solid	01 solid	Thermal
		02 solid	Thermal
		03 solid	Thermal
		04 solid	Thermal
		05 solid convection	Thermal
	02 surface	06 surface convection	Thermal

Figure 2-12: Names and types of studies completed in this chapter.

3: L bracket

Topics covered

- ❑ Heat transfer by conduction
- ❑ Use of 2D models
- ❑ Singularities in thermal problems

Project description

We will conduct a thermal analysis of a simple model to study heat flux singularities. Demonstration of heat flux singularities requires the use of very small elements, consequently models are large and expensive to solve. However, our model lends itself toward a 2D representation. Using 2D thermal elements, we will demonstrate a heat flux singularity at a fraction of the computational effort that would be required if a 3D analysis were performed.

Open model L BRACKET TH shown in Figure 3-1; make sure it is in the *01 sharp edge* configuration.

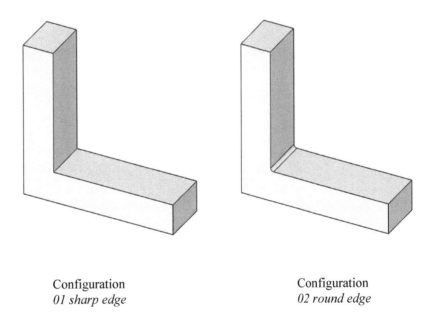

Configuration
01 sharp edge

Configuration
02 round edge

Figure 3-1: Model L BRACKET TH comes in two configurations shown above.

Demonstrating heat flux singularities requires the 01 sharp edge configuration.

Create thermal study *01 sharp* using 2D Simplification (Figure 3-2).

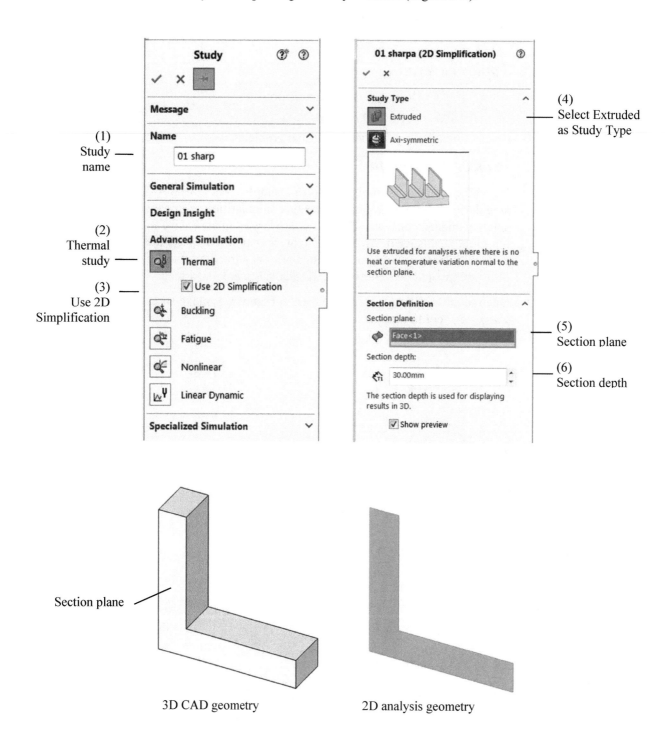

Figure 3-2: Model L BRACKET TH can be represented by a 2D section.

Section depth is the thickness of the CAD model; the side face is used as a section plane.

To demonstrate the heat flux singularity we must, of course, define heat flow in the model. The easiest way to do that is to define prescribed temperatures as boundary conditions as shown in Figure 3-3.

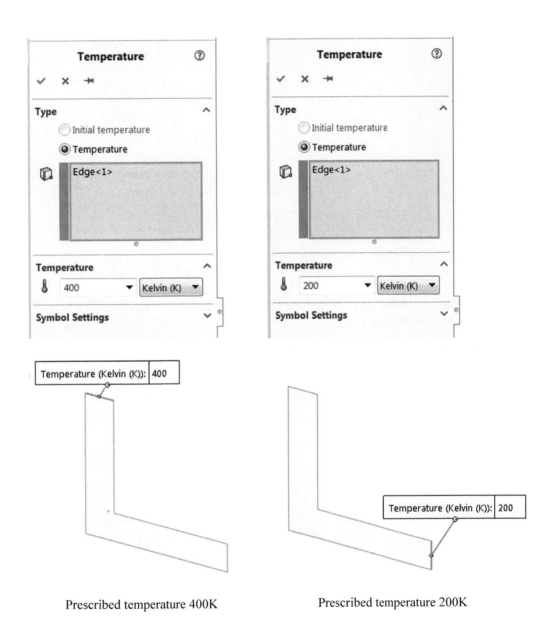

Prescribed temperature 400K   Prescribed temperature 200K

Figure 3-3: Definition of prescribed temperatures in 2D thermal model; top edge: 400K, right edge: 200K.

The temperature 400K is defined on the top horizontal edge; the temperature 200K is defined on the right vertical edge.

Mesh the model with an element size of 2mm, run the solution and display a resultant **Heat Flux** plot (Figure 3-4).

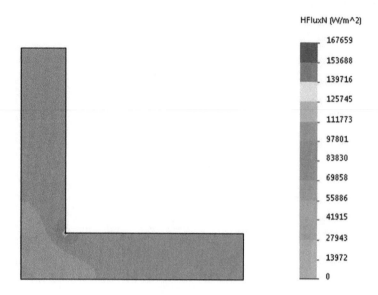

Figure 3-4: Heat flux results obtained with 2mm element size show the maximum heat flux 168kW/m^2.

Review this plot using vector display.

Notice that the location of the maximum heat flux coincides with the sharp corner.

Copy study *01 sharp* into study *02 sharp*; add a mesh control to the corner (Figure 3-5):

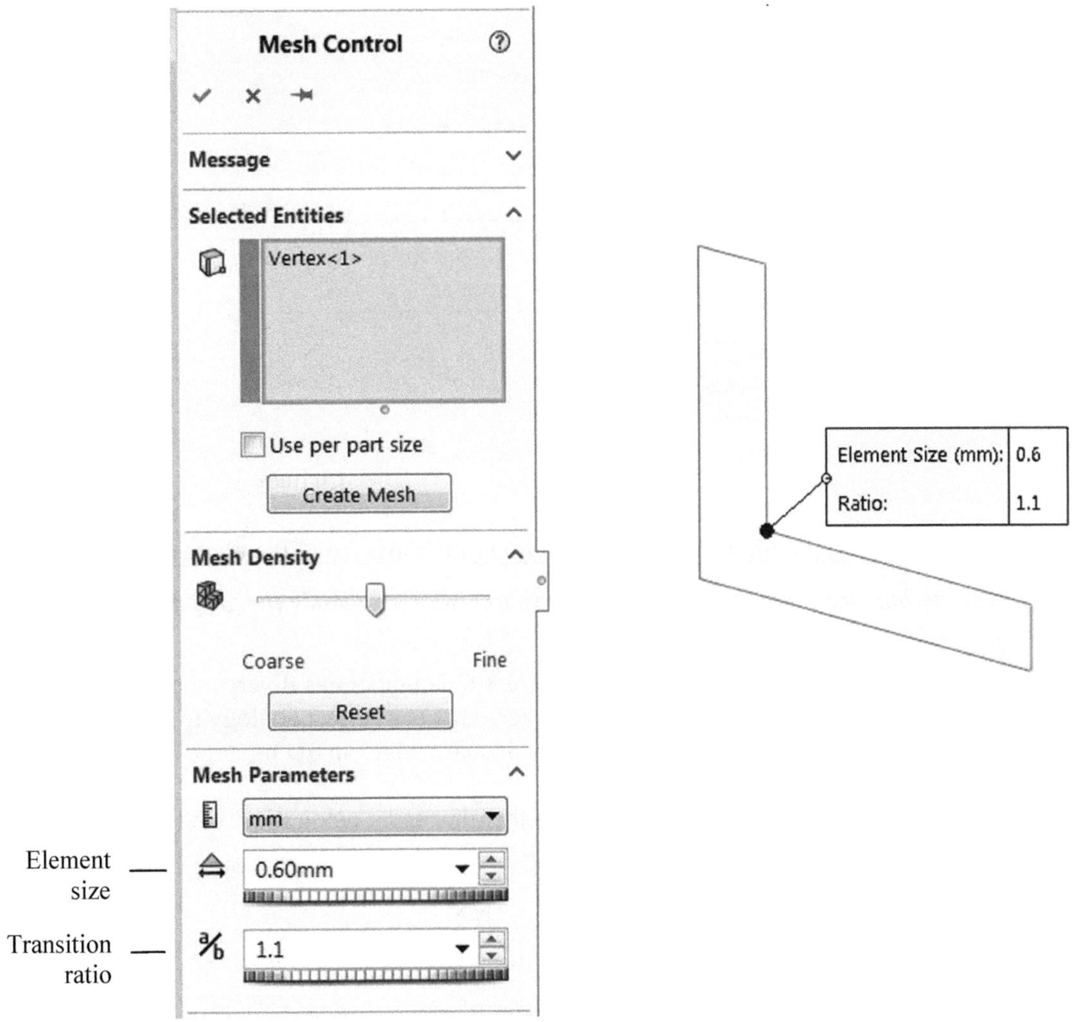

Figure 3-5: Mesh control in the sharp corner in study *02 sharp*.

Use an element size of 0.6mm, and a transition ratio of 1.1 to ensure a smooth transition from dense elements in the corner to coarse elements (2mm) everywhere else in the model.

Run the analysis then repeat it using progressively smaller elements in the corner; study *03 sharp*: 0.4mm, study *04 sharp*: 0.2mm and study *05 sharp*: 0.1mm. Collect heat flux results in the corner (this will also be the maximum heat flux) from all studies and present them in a graph showing the maximum heat flux as a function of 1/h where h is the element size in the corner (Figure 3-6).

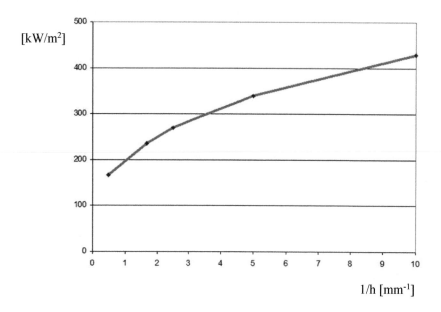

Figure 3-6: Maximum heat flux as a function of the inverse of the element size h.

Five points are connected by straight lines to visually enhance the graph.

The summary of results presented in Figure 3-6 demonstrates divergence of heat flux results in the presence of a sharp corner. This is in direct analogy to stress divergence in sharp corners in 2D models or sharp edges in 3D models.

Divergence of heat flux results will be further discussed using the HEATING DUCTS model in chapters 5 and 6.

Summary of studies completed

Model	Configuration	Study Name	Study Type
L BRACKET TH.sldprt	*01 sharp edge*	*01 sharp*	Thermal
		02 sharp	Thermal
		03 sharp	Thermal
		04 sharp	Thermal
		05 sharp	Thermal
	02 round edge		

Figure 3-7: Names and types of studies completed in this chapter.

Notes:

4: Thermal analysis of a round bar

Topics covered

- ❑ Heat transfer by conduction
- ❑ Thermal conductivity
- ❑ Heat transfer by convection
- ❑ Convection boundary conditions
- ❑ Thermal resistance
- ❑ Prescribed temperature boundary conditions
- ❑ Heat power
- ❑ Heat flux

Project description

A round steel rod is represented by STEEL ROD part model; it has one end in boiling water at 100°C and the other end in melting ice (0°C). The cylindrical face is insulated; no heat is exchanged through that face (Figure 4-1).

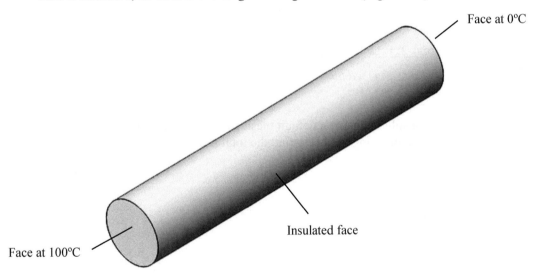

Face at 0°C

Insulated face

Face at 100°C

Figure 4-1: The STEEL ROD model with prescribed temperatures as boundary conditions.

If no thermal boundary conditions are defined on a face, it means that no heat is exchanged through that face; the face is insulated.

Our objective is to find the total heat transferred by the rod from the hot to cold side. Notice that conduction is the only heat transfer mechanism present in the model.

Create a thermal study *01 conduction* and define prescribed temperatures as boundary conditions as shown is Figure 4-2.

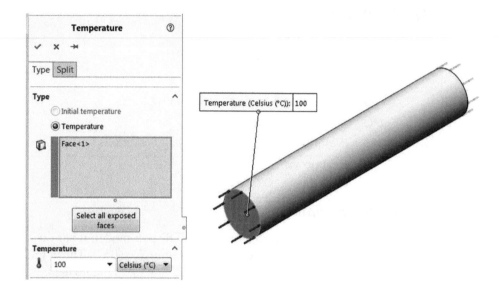

Figure 4-2: Prescribed temperature 100°C defined on the hot side of the rod.

Define a prescribed temperature of 0°C on the other side.

Obtain a solution using a default mesh size and review the **Temperature** plot. Follow steps in Figure 4-3 to find the total heat traveling through the rod.

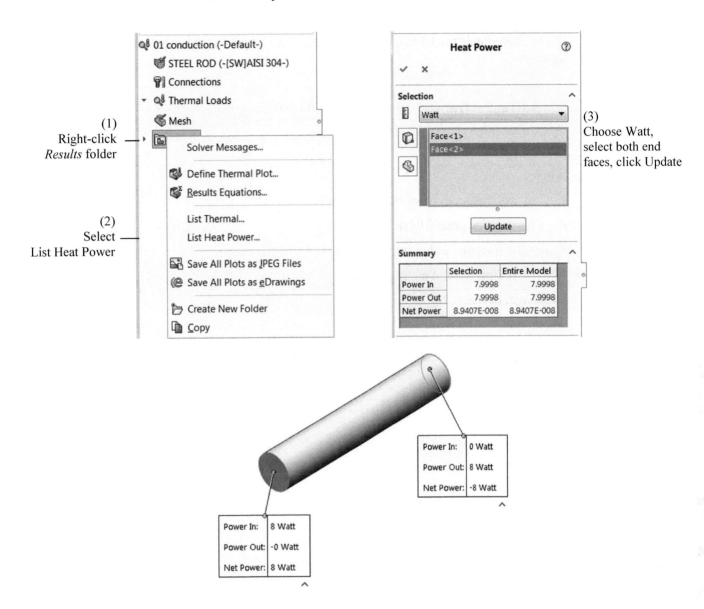

Figure 4-3: Listing Heat Power.

Heat power entering the model is 8W; heat power exiting the model is 8W.

We use this simple model to compare our heat power result to the analytical solution:

$$Q = \frac{kA(T_{HOT} - T_{COLD})}{L} = \frac{16 \times 0.001 \times (100 - 0)}{0.2} = 8W$$

Where:

k – Thermal conductivity of steel: 16W/m/K

A – Cross section area: $0.001m^3$

T_{HOT} – Temperature of boiling water: 100°C

T_{COLD} – Temperature of melting ice: 0°C

L – Length of rod: 0.2m

To complete the review of the results of study *01 conduction*, create a **Heat Flux** plot presented as vectors (Figure 4-4).

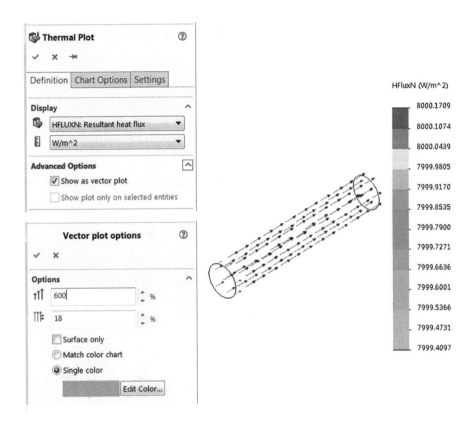

Figure 4-4: The resultant heat flux presented as vector plot with Vector plot options shown above.

Heat flux is uniform throughout the model as shown by the numbers at the color legend. The only reason for minute differences is a discretization error introduced by meshing.

Next, we introduce heat transfer through convection. Copy *study 01 conduction* into *02 convection* and delete the prescribed temperature on the cold side. We want to replace this prescribed temperature boundary condition with a convection boundary condition that will give the same temperature distribution along the length of the rod and the same heat flow 8W. What should be the convection coefficient?

First, let us consider what ambient temperature should be defined. We want the cold end of the rod to be at 0°C (273K). If we defined the ambient temperature equal to the previously used prescribed temperature 0°C (273K), then the convection coefficient would have to be infinitely large. The ambient temperature must be lower than 0°C. The lower the ambient temperature, the lower will be the required convection coefficient while 8W of heat power travels through the rod. Let us assume the ambient temperature is -10 °C.

Heat transferred by convection can be found using this equation:

$$Q = hA(T_S - T_F)$$

Where:

Q – Heat power (8W) - given

h – Unknown convection coefficient [W/m²/K]

A – Area participating in the convective heat transfer (0.001m²) - given

T_S – Temperature of the surface (273 K) - given

T_F – Temperature of the surrounding fluid (263K) - assumed

From the above equation, the required convection coefficient is:

$$h = \frac{Q}{A(T_S - T_F)} = \frac{8}{0.001 \times 10} = 800 \frac{W}{m^2 K}$$

Using this result, define the convection coefficient as shown in Figure 4-5.

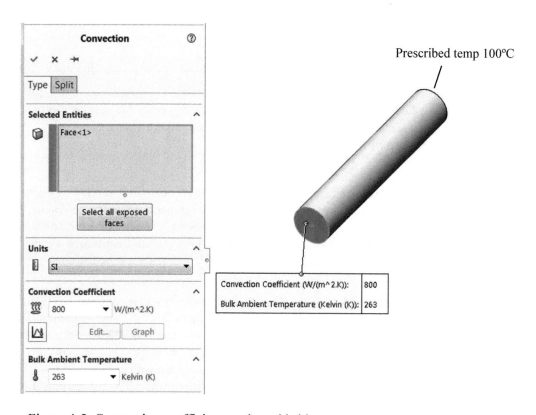

Figure 4-5: Convection coefficient on the cold side.

The convection coefficient replaces the previously used prescribed temperature boundary condition. Keep the prescribed temperature 100°C on the opposite side.

Obtain the solution and review the temperature plot shown in Figure 4-6.

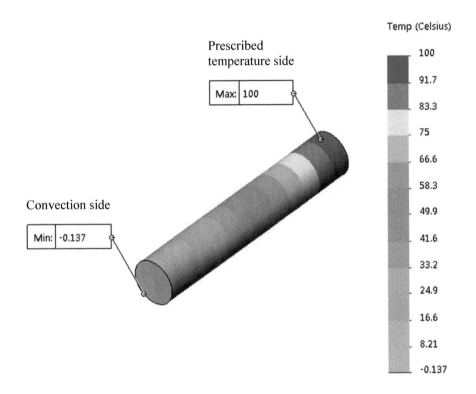

Figure 4-6: Temperature plot; callouts indicate the minimum and the maximum temperature in the rod.

The maximum temperature is equal to the prescribed temperature 100°C; the minimum temperature is close to 0°C.

Notice linear distribution of temperature along the length of the rod; this is because no heat is exchanged through the cylindrical face.

Find the heat power as shown in Figure 4-7.

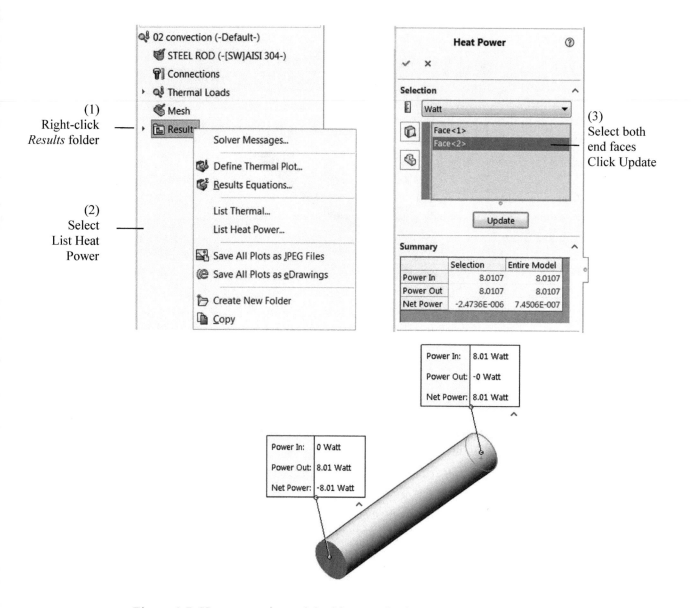

Figure 4-7: Heat power in model with prescribed temperature on the hot side and convection on the cold side.

Compare this to the results shown in Figure 4-3.

As Figure 4-6 shows, the convection coefficient of 800W/m²/K combined with the temperature gradient between the face and surrounding fluid is equivalent to the prescribed temperature of 273K. From a practical point of view we should notice that the convection coefficient of 800 W/m²/K is quite high; however, it would be possible to get that convection between metal and moving water with melting ice.

Open assembly TWO RODS; the assembly consists of two identical copper rods 100mm long glued with a layer of epoxy resin. The thickness of the glue is 0.5mm; not a good design but makes for an interesting exercise (Figure 4-8).

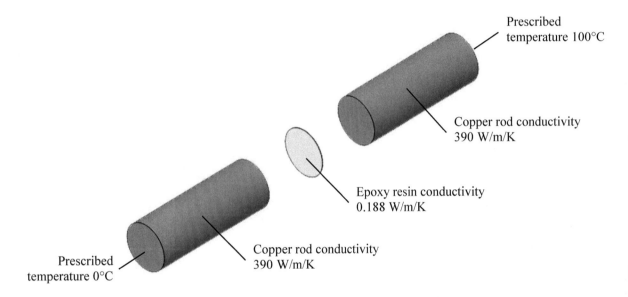

Prescribed temperature 100°C

Copper rod conductivity 390 W/m/K

Epoxy resin conductivity 0.188 W/m/K

Copper rod conductivity 390 W/m/K

Prescribed temperature 0°C

Figure 4-8: The TWO RODS model features two copper rods glued by epoxy glue. Prescribed temperatures on both ends are shown.

The rod material (copper) and glue material (epoxy resin) have vastly different thermal conductivities. Assembly is shown in an exploded view.

Our objective here is to study the effect of thermal resistance caused by the layer of epoxy. Even though the layer of epoxy is very thin, it has a major effect on the heat transfer through the assembly because of its low thermal conductivity. The thinness of the epoxy layer will also cause significant meshing difficulties.

Make sure the assembly is in the *01glue* configuration and create a thermal study called *01 glue*. To ensure a correct aspect ratio of elements meshing the glue layer, apply a **Mesh Control** to the glue component as shown in Figure 4-9.

Select glue from the fly-out menu

Element size 0.5mm

Ratio 1.2

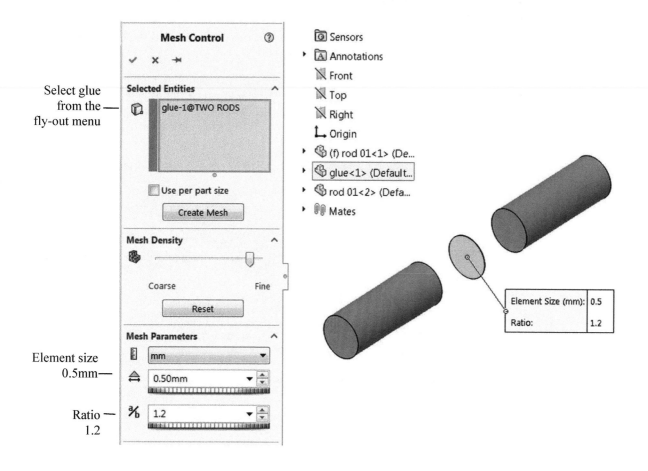

Figure 4-9: Mesh Control applied to glue assembly component.

A low a/b ratio assures a smooth transition between large elements that mesh the rods, and small elements that mesh the glue. Assembly shown in exploded view.

The mesh control shown in Figure 4-8 produces one layer of elements in the glue. Second order elements are capable of modeling the second order temperature distribution. Given the expected linear temperature drop across the epoxy layer, one layer of elements will suffice.

Mesh the assembly with a default element size. Apply prescribed temperatures identical to the STEEL ROD exercise: 100°C on one end, 0°C on the other end (Figure 4-8).

Run the solution and review the **Temperature** plot. Notice the temperature drop of 8°C over the distance of 100mm in the bar. Compare this to a drop of 84°C over the distance of 0.5mm in glue (Figure 4-10).

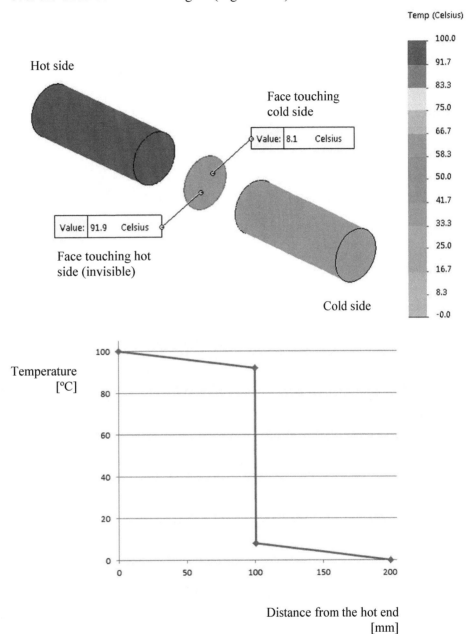

Figure 4-10: Temperature drop along the length of the assembly.

A rapid temperature drop is caused by a low thermal conductivity of the epoxy glue.

Total heat flow in the assembly can be displayed as shown in Figure 4-11.

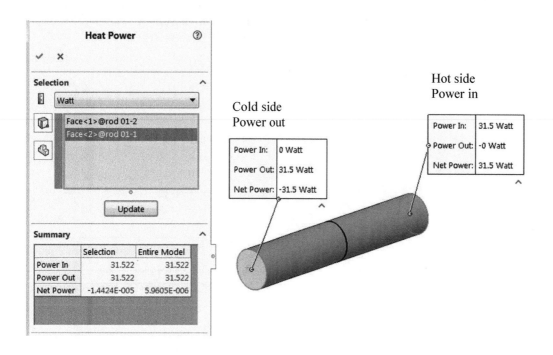

Figure 4-11: Total heat flow through the assembly.

Since there is no heat loss or gain along the length of the assembly, the magnitude of the entering heat power 31.5W equals the magnitude of the exiting heat power 31.5W.

The thermal resistance of epoxy causes a temperature a drop of 83°C over the distance of 0.5mm. Heat flowing through the epoxy is 31.5W. The cross sectional area of the epoxy is 1000mm². Knowing the above we may calculate the thermal resistance of the epoxy layer:

$$R = \frac{\Delta T A}{P} = \frac{83 \times 0.001}{31.5} = 0.0026 \frac{Km^2}{W}$$

Where:

R – Distributed (per area) thermal resistance Km²/W

ΔT – Temperature drop across the layer 83K

A – Cross sectional area: 0.001m³

P – Heat flow (heat power) through the layer

The glue part of the assembly is very small in comparison to the rods, yet it causes a major obstacle in heat flow because of its poor heat conductivity. It also poses major modeling difficulties; we had to apply a mesh control that increased the number of elements in the model very significantly.

The objective of this exercise is to introduce the concept of the layer of thermal resistance. We saw that explicit modeling of the layer of thermal resistance is difficult and gets even more so when the layer gets very thin. Fortunately, there is a better way of modeling the layer of thermal resistance.

Stay with the TWO RODS assembly but switch to the *02 no glue* configuration. The assembly now consists of two parts only. Create a thermal study *02 thermal resistance*. The layer of thermal resistance will be modeled as a **Contact Condition**. Follow Figure 4-12 to define a **Thermal Resistance Contact Condition**.

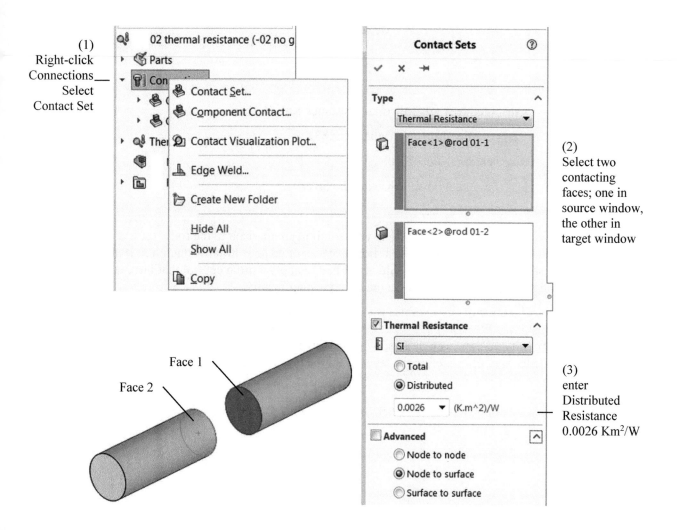

(1)
Right-click
Connections
Select
Contact Set

(2)
Select two contacting faces; one in source window, the other in target window

(3)
enter Distributed Resistance 0.0026 Km^2/W

Figure 4-12: Definition of a Thermal Resistance Contact Condition.

Thermal Resistance is the only type of contact condition available in a thermal analysis. Use exploded view.

The value of thermal resistance, 0.0026 Km^2/W, comes from a calculation based on the results of the previous study where the layer of thermal resistance was modeled explicitly.

Mesh the assembly with the default element size; no mesh control is needed. Obtain a solution and review the **Temperature** plot.

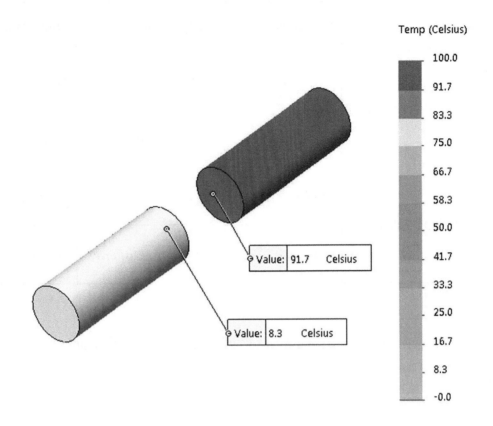

Figure 4-13: Temperature results in the model with a layer of thermal resistance.

The temperature drop across the layer of thermal resistance is 83°C.

The results of study *01 glue* (Figure 4-10) and study *02 thermal resistance* (Figure 4-13) are practically the same. The layer of epoxy glue produces the same temperature drop when modeled either explicitly or as a layer of thermal resistance.

The use of **Thermal resistance Contact Conditions** allows for a very significant simplification of the model, and a faster solution time compared to explicit modeling, as long as the value of thermal resistance is known. If it is not known, it can be found using simple models such as TWO RODS in the *01 glue* configuration. The analysis of real design problems can then be done using **Thermal Resistance**.

Summary of studies completed

Model	Configuration	Study Name	Study Type
STEEL ROD.sldprt	*Default*	*01 conduction*	Thermal
		02 convection	Thermal
TWO RODS.sldasm	*01 glue*	*01 glue*	Thermal
	02 no glue	*02 thermal resistance*	Thermal

Figure 4-14: Names and types of studies completed in this chapter.

5: Floor heating duct – part 1

Topics covered

- ❑ Heat transfer by conduction
- ❑ Prescribed temperature boundary conditions
- ❑ Heat power
- ❑ Heat flux
- ❑ Heat flux singularities
- ❑ Analogies between structural and thermal analysis

Project description

A ceramic floor panel used to heat the room is modeled in assembly HEATING DUCTS (Figure 5-1). Hot air flows in the ducts transferring heat to the panel. Heat is dissipated by the panel's top faces. We assume that the side and bottom faces are insulated, therefore no thermal boundary conditions are defined there: no prescribed temperatures, no convection coefficients and no radiation coefficients.

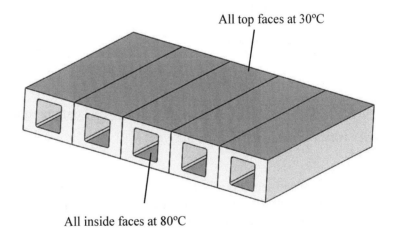

All top faces at 30°C

All inside faces at 80°C

Figure 5-1: HEATING DUCTS assembly model of a ceramic floor panel.

The assembly consists of five identical parts. The model is shown in 01 CAD configuration.

The temperatures shown in Figure 5-1 are prescribed temperatures; we will investigate heat flow induced by these thermal boundary conditions. Side faces and bottom faces do not have any prescribed temperature defined but that does not mean that they will not have any temperature. Temperature there will establish itself as a result of heat flow induced by prescribed temperatures shown in Figure 5-1.

We will perform a thermal analysis of this floor panel to find out how much heat is transferred to the room. Because of repeatable geometry and no thermal boundary conditions on either side, the model can be simplified to just one segment. Also, because there is no temperature gradient along the length of ducts (uniform prescribed temperature is defined) the analysis can be further simplified by using a 2D model. A 2D model will also make it easy to demonstrate the heat flux singularity in the presence of sharp edges.

Switch to the *02 Analysis Part1* configuration to review the model (Figure 5-2).

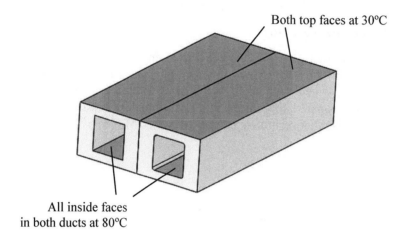

Both top faces at 30°C

All inside faces
in both ducts at 80°C

Figure 5-2: HEATING DUCTS assembly in *02 FEA* configuration.

Notice round edges in the duct on the right.

Why do we use two segments for analysis even though only one segment could be used? We do that to study the effect of sharp re-entrant edges on heat flux; notice that the left segment has sharp edges; the right one has rounded edges.

The analyzed segment consisting of two ducts is part of a larger assembly. No heat flows through the side faces and bottom; therefore, no boundary conditions are defined there. The side and bottom faces are treated as if they were insulated.

Procedure

Make sure the model is in configuration *02 Analysis Part 1* and define a thermal study *01 2D* using **2D Simplification**. Follow the steps shown in Figure 5-3.

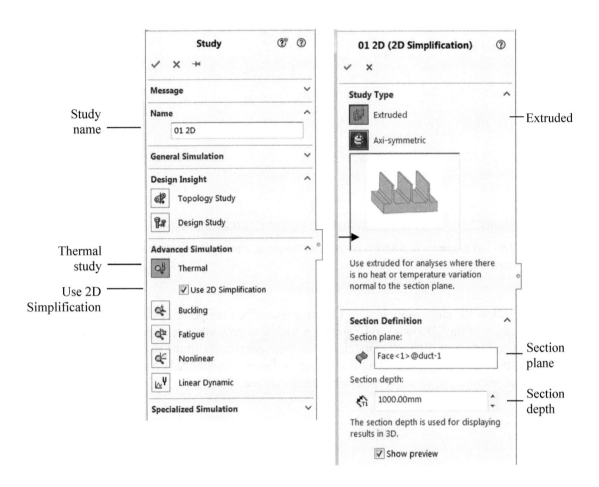

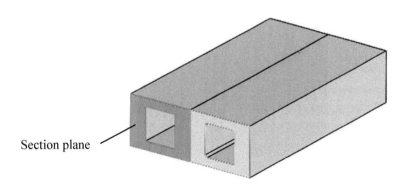

Section plane

Figure 5-3: Thermal study definition using 2D Simplification.

Section depth is equal to the length of the floor panel 1000mm; this can be measured in the CAD model.

Geometry ready to be meshed with 2D elements is shown in Figure 5-4.

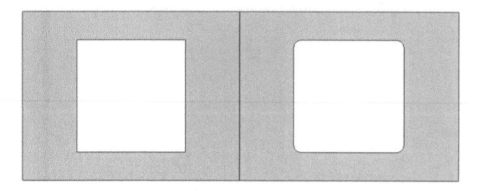

Figure 5-4: 2D geometry ready to be meshed with 2D elements.

The left segment has sharp re-entrant corners; the right segment has round corners.

Follow the steps in Figure 5-5 to define prescribed temperatures on the inside edges of the 2D model and Figure 5-6 to define prescribed temperatures on the outside edges of the 2D model. Notice that contact conditions remain as default **Bonded** meaning that there is no layer of thermal resistance between two segments. No changes to contact conditions are required.

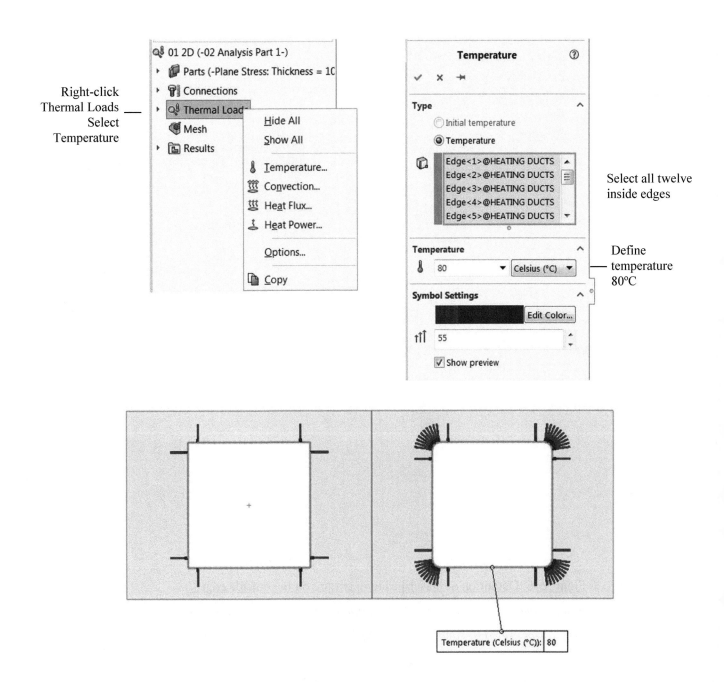

Figure 5-5: Definition of prescribed temperature on the inside edges.

A prescribed temperature of 80°C is applied to twelve inside edges: four edges in the left segment and eight edges in the right segment. Prescribed temperature symbols have been made smaller.

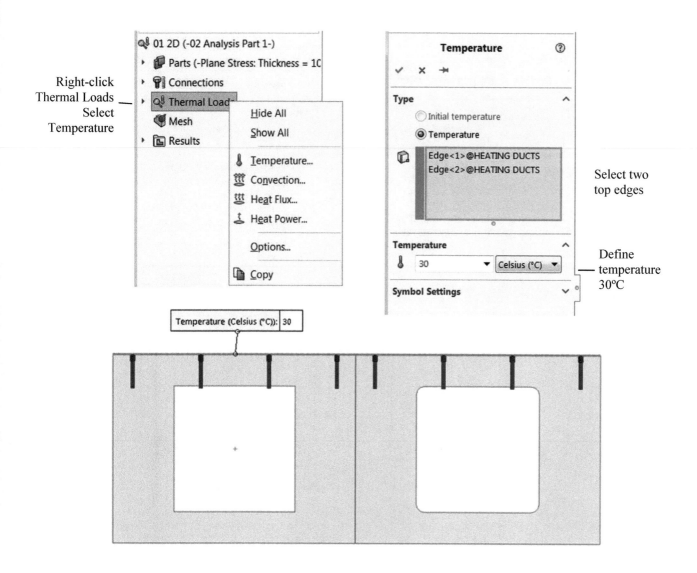

Figure 5-6: Definition of prescribed temperature on the outside edges.

A prescribed temperature of 30°C is applied to the two top outside edges.

Symbols of previously defined prescribed temperatures are not shown.

Having defined prescribed temperatures on the inside and outside edges we have established a mechanism of heat transfer. Now, mesh the model with a default element size to create a mesh shown in Figure 5-7.

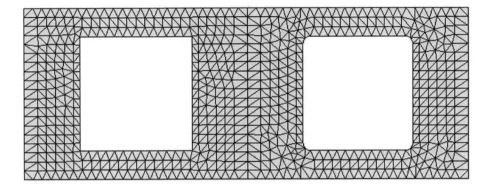

Figure 5-7: Model meshed with default element size.

Notice the high turn angle in the elements meshing round corners.

Obtain a solution noticing almost instantaneous results; 2D thermal models solve very fast because of a low number of elements due to the 2D formulation and to the fact that there is only one degree of freedom (temperature) per node. Review temperature results as shown in Figure 5-8.

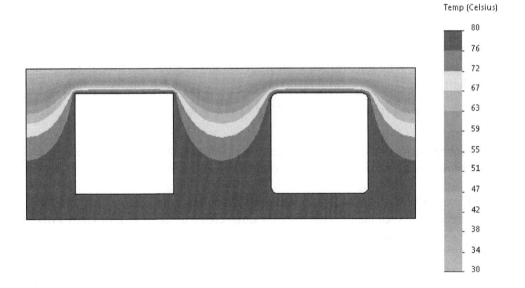

Temp (Celsius)

80
76
72
67
63
59
55
51
47
42
38
34
30

Figure 5-8: Temperature results from study *2D thermal 01*.

Temperature is a scalar value; temperature results can be presented only as a fringe plot.

As Figure 5-8 illustrates, there is little difference between sharp and round corners. The temperature on the inside edges is equal to 80°C; the temperature on the outside edges is equal to 30°C. Everywhere else, a steady state temperature establishes itself as a result of heat flow from hot (80°C) to cold (30°C) edges. Now, create Heat Flux plot as shown in Figure 5-9.

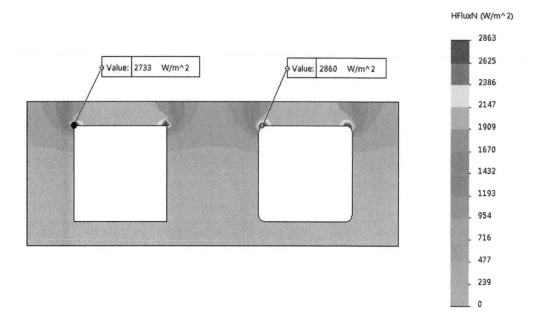

Figure 5-9: Heat Flux results from study *2D thermal 01*. The probed result in the sharp corner is $2733 W/m^2$; the probed result in the round corner is $2860 W/m^2$.

The fringe plot is used to present Heat Flux results.

Figure 5-9 shows a small difference in **Heat Flux** between the sharp and round corners, but this will change once we use a more refined mesh.

Heat Flux is a vector value; fringe plots show the magnitude but not the direction. To see the direction of heat flow, create a vector plot of **Heat Flux** as shown in Figure 5-10.

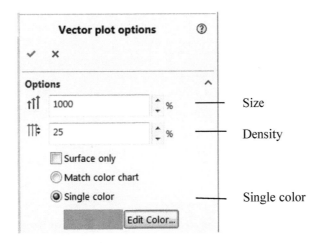

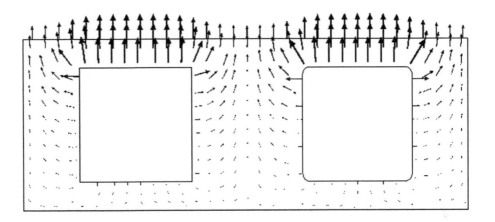

Figure 5-10: Heat Flux results from study *2D thermal 01*.

The vector plot may be modified using Vector plot options shown on the left. Color information is suppressed. Experiment with different Vector plot options.

Figure 5-10 makes it clear that heat enters and escapes the model only through the edges where prescribed temperatures have been defined. No heat enters or escapes the model through the bottom or sides.

Copy study *01 2D* into *02 2D* and add mesh controls as shown in Figure 5-11.

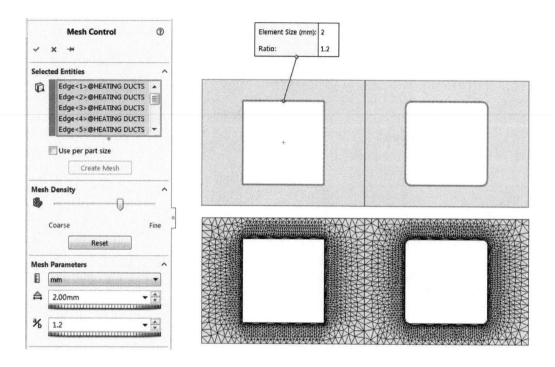

Figure 5-11: Mesh controls and mesh in study *02 2D.*

An element size of 2mm is defined on all edges of both holes. A transition ratio of 1.2 creates a large transition zone.

Using a small element size as shown in Figure 5-11 will demonstrate a very important difference between heat flow in the presence of a sharp and round corner. Obtain the solution of *02 2D* and review the **Temperature** and **Heat Flux** results shown in Figure 5-12 and Figure 5-13.

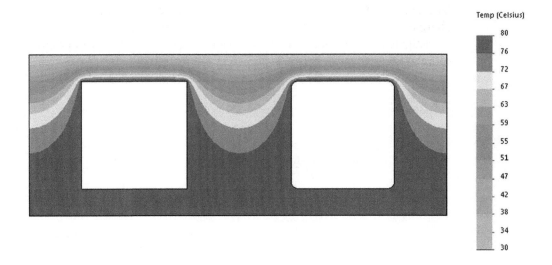

Figure 5-12: Temperature results from study 02 *2D*.

These results are indistinguishable from the results of study 2D thermal 01.

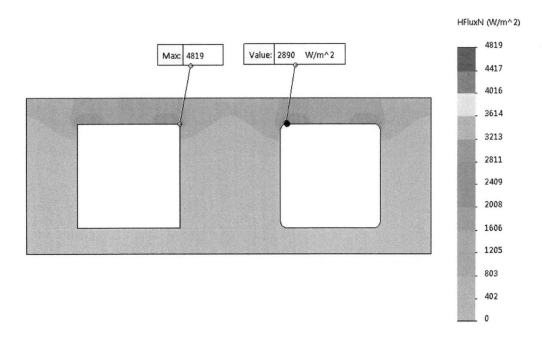

Figure 5-13: Heat Flux results from study 02 *2D*. The maximum value coincides with the sharp corner and reads 4819 W/m^2. A probed result in the round corner shows 2890W/m^2.

Due to high Heat Flux gradients in sharp corners and in round corners, the magnitudes displayed by probing strongly depend on the location of probing. You may see results slightly different from what is shown in the illustrations.

The small elements used in study *02 2D* reveal a heat flux concentration in the sharp corners. The concentration was previously masked by large elements. To investigate this concentration, we will run three more studies with progressively more aggressive mesh biases as shown in Figure 5-14.

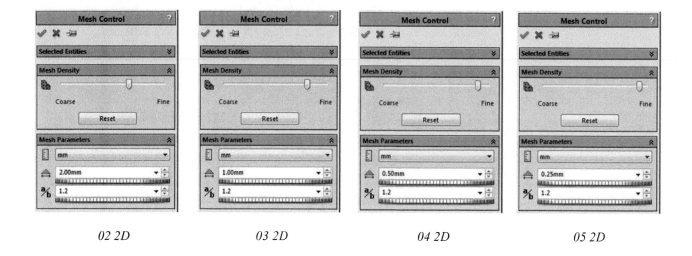

02 2D   03 2D   04 2D   05 2D

Figure 5-14: Mesh control in studies *02 2D* through *05 2D*.

There is no Mesh Control in study 01 2D. The global element size (12mm) is the same in all studies.

The summary of results is shown in Figure 5-15.

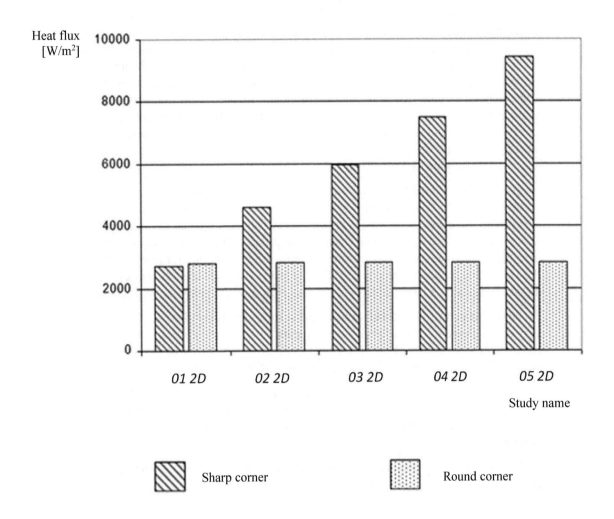

Figure 5-15: Summary of Heat Flux results in five studies.

The Heat Flux magnitude diverges with mesh refinement in the sharp corner while mesh refinement has no noticeable effect on Heat Flux in the round corner.

Divergent heat flux results in a sharp corner are caused by a singularity. The mathematical model has no heat flux solution in a sharp corner; a mathematician would say that the solution is singular there; an engineer would say that heat flux is infinite in a sharp re-entrant corner. Meshing introduces a discretization error, which masks the singularity and may give an appearance of correct results especially when large elements are used as in study *01 2D*. Mesh refinement reveals the singularity demonstrating divergent results (Figure 5-15). Numerical values of heat flux in sharp and round corners seen in study *01 2D* are close by coincidence. We can't say which study provides "the best" heat flux results in sharp corners because all these results are meaningless. Heat flux results in locations coinciding with sharp corners (or edges in case of 3D models) are entirely dependent on the choice of discretization and by manipulating the mesh size we can produce as high a result as we want!

Notice that sharp corners have no effect on temperature results. Sharp re-entrant corners (in 2D models) or sharp re-entrant edges (in 3D models) may be acceptable if the data of interest is temperature. They can also be acceptable if heat flux results are sought away from the singularity. However, if the maximum heat flux is the data of interest, rounds must be modeled no matter how small they are. Notice that a sharp (zero-radius) corner or edge is a mathematical abstract; in reality a round is always present even though it may be very small.

To demonstrate the divergence of heat flux results (singularity) we had to use very small elements. It was possible to execute a number of studies in a short time because we used a 2D model. Divergent heat flux can also be demonstrated with a 3D model but at a higher computational effort.

We will now look at the heat power of the floor panel. Heat power is not sensitive to mesh refinement or heat flux singularities so we may use any of the five studies created so far. Heat power is analogous to the reaction force in a structural analysis.

To list heat power for two segments we use results from study *01 2D* and follow the steps shown in Figure 5-16.

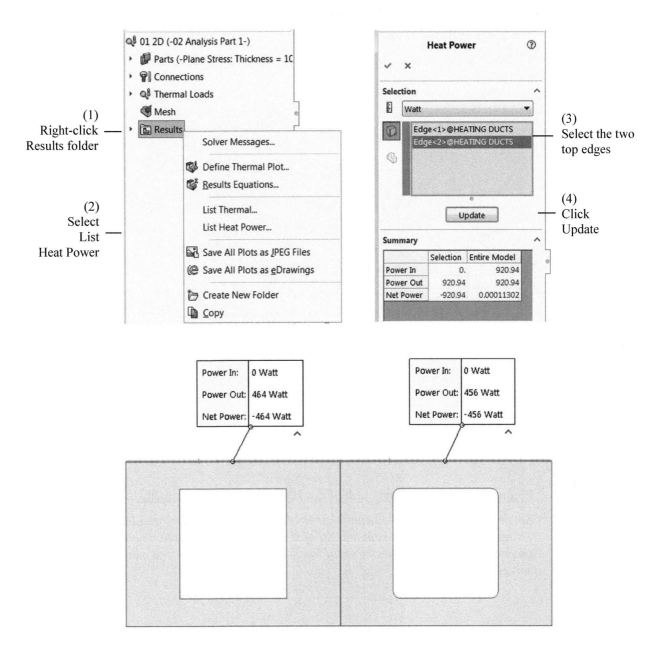

Figure 5-16: The heat power of the segment with sharp corners is 464W; the heat power of the segment with round corners is 4456W. The total heat power of the two segments is 921W.

The difference in heat power is due to the different amount of conducting material.

To wrap up the analysis of the floor panel with prescribed temperatures used as boundary conditions we will create one more study: *06 3D* where a 3D model will be analyzed to find the total heat power.

Use a standard mesh with element size 10mm to create mesh as shown in Figure 5-17.

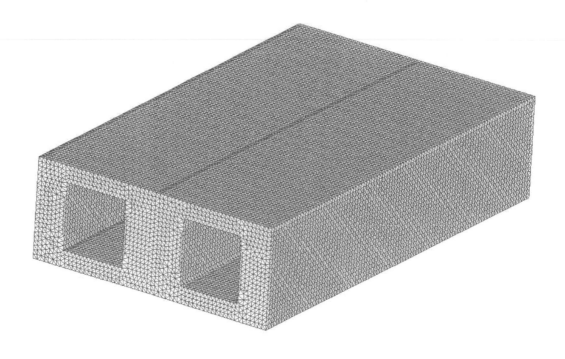

Figure 5-17: Solid element mesh of the 3D model with two segments.

Small elements are used to avoid high element distortions in the round corners. Using global mesh refinement instead of mesh controls is acceptable because thermal elements have one degree of freedom per node and a large model still solves fast.

Heat Power results are shown in Figure 5-18.

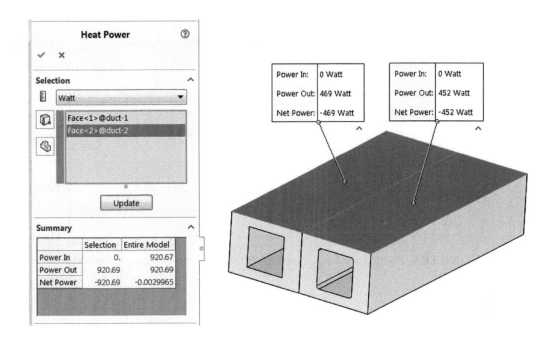

Figure 5-18: The heat power of the segment with sharp edges is 469W; the heat power of the segment with round edges is 452W. The total heat power of the two segments is 922W.

Small differences in results between the 2D model (Figure 5-16) and the 3D model (Figure 5-18) are caused by discretization errors; discretization errors are different between 2D and 3D models.

Summary of studies completed

Model	Configuration	Study Name	Study Type
HEATING DUCTS.sldasm	01 CAD		
	02 Analysis Part 1	01 2D	Thermal
		02 2D	Thermal
		03 2D	Thermal
		04 2D	Thermal
		05 2D	Thermal
		06 3D	Thermal
	03 Analysis Part 2		

Figure 5-19: Names and types of studies completed in this chapter.

6: Floor heating duct – part 2

Topics covered

- ❏ Heat transfer by convection
- ❏ Free and forced convection
- ❏ Convection coefficient
- ❏ Ambient (bulk) temperature

Project description

We have completed part 1 of the floor heat panel analysis modeling heat flow induced by prescribed temperatures: 80°C on the hot side and 30°C on the cold side. Now, we will take this analysis one step further.

Assume that we have taken the following temperature measurements: air inside: 85 °C, inside faces: 80°C, air outside: 17°C, outside face: 30°C. Notice that the face temperatures are the prescribed temperatures from part 1 of this analysis.

Given the above temperature information we want to find out what convection conditions must be present on the hot and on the cold side to keep these steady state temperatures. For that we use the same assembly model HEATING DUCTS and switch to configuration *03 Analysis Part2*; this configuration has only one segment (Figure 6-1).

Air outside 17°C (290K)
Outside face 30°C (303K)

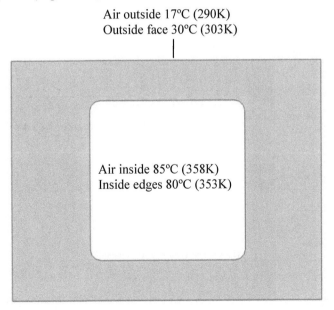

Air inside 85°C (358K)
Inside edges 80°C (353K)

Figure 6-1: HEATING DUCTS assembly model in *03 Analysis Part3* configuration.

A 2D Simplification of one segment with round edges is used. The outside wall is represented by the top edge.

The bottom outside face is assumed to be insulated and side faces don't exchange heat because this segment is a part of the larger assembly; therefore, no thermal boundary conditions are defined on these faces. We will proceed in two steps using the 2D model; refer to Figure 6-1 for the location of inside and outside edges.

Step 1 (study *07 2D*):

Thermal boundary conditions on the inside edges are prescribed temperatures of 80°C; this is unchanged from part 1. The convection coefficient on the outside edge will be changed in steps from 10W/m^2/K to 1000W/m^2/K to determine at what value an average outside edge temperature of 30°C is produced.

Step 2 (study *08 2D*):

The convection condition on the outside edge will be the one found in step 1. The convection coefficient on the inside edges will change in steps from 10W/m^2/K to 1000W/m^2/K to determine at what value an average inside edge temperature of 80°C is produced.

Once the convection coefficients on both sides have been found, there won't be a need to define prescribed temperatures, and heat flow will be modeled more realistically.

To prepare study *07 2D,* use **2D simplification**. Define **Convection** as shown in Figure 6-2. Define a prescribed **Temperature** of 80°C on the inside edges as in part 1 of this exercise.

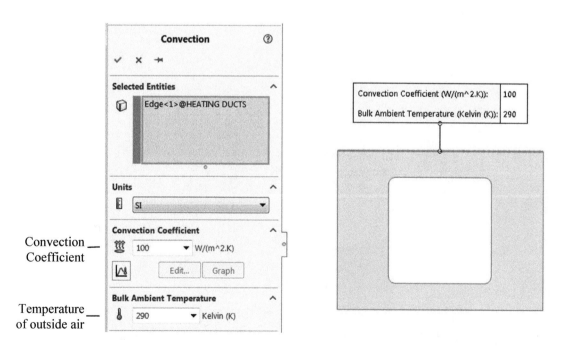

Figure 6-2: Convection coefficient on the outside edge.

A sample value of 100W/m^2/K is shown.

Using default element size run study *07 2D* for the following values of convection coefficients on the outside edge: 10, 50, 100, 200, 300, 500, 1000; you may create separate studies. For each value, find the average temperature on the outside edge as shown in Figure 6-3.

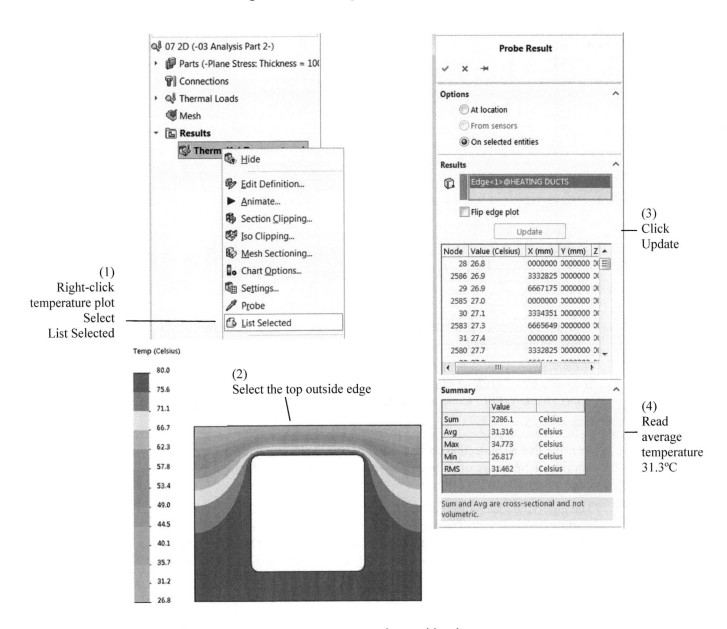

Figure 6-3: Average temperature on the outside edge.

With a convection coefficient of 100W/m²/K, the average temperature of the outside edge is 31.3°C.

The summary of all runs for convection coefficients on the outside edge of 10, 50, 100, 200, 300, 500, and 1000, is shown in Figure 6-4.

Average temp.
outside edge
[°C]

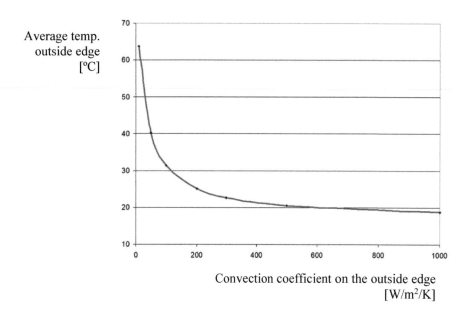

Convection coefficient on the outside edge
[W/m²/K]

Figure 6-4: Average temperature on the top outside edge as a function of convection coefficient.

A convection coefficient of 100W/m²/K in combination with an ambient temperature of 290K results in an average temperature of the outside edge of approximately 30°C. Compare this with Figure 6-3.

We find that a convection coefficient of 100W/m²/K in combination with an ambient temperature of 290K results in an average temperature of the outside edge of approximately 30°C.

Having found the convection coefficient on the outside edge we proceed to finding convection coefficient on the inside edges. Copy study *07 2D* into *08 2D* and delete the prescribed temperature on the inside edges. Define convection on the outside edge with the convection coefficient 100W/m²/K and the ambient temperature 290K and keep it unchanged throughout the analysis.

Run study *08 2D* for the following values of convection coefficients on the inside edges: 10, 50, 100, 200, 300, 500, and 1000; refer to Figure 6-5.

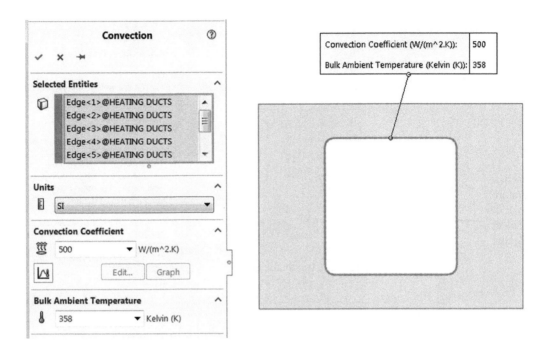

Figure 6-5: Convection coefficient on the inside edges.

A sample value of 500W/m²/K is shown in the Convection window.

For each value of convection coefficient find the average temperature on the inside edge as shown in Figure 6-6.

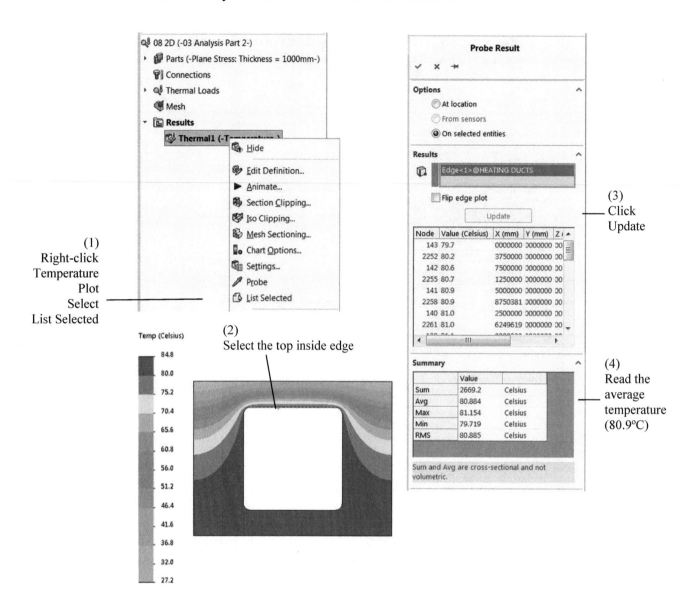

Figure 6-6: Average temperature on the inside edge.

A convection coefficient of 500W/m²/K gives the average temperature of 80.9°C on the inside edge.

A summary of all runs for convection coefficients on the inside edges of 10, 50, 100, 200, 300, 500, and 1000, is shown in Figure 6-7.

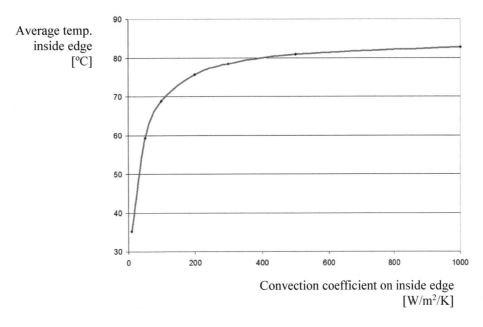

Figure 6-7: Average temperature on the top inside edge as a function of convection coefficient.

A convection coefficient of 400W/m²/K gives an average temperature close to 80°C on the inside edge. Compare this with Figure 6-6 showing results for 500W/m²/K.

Run study *08 2D* one more time with a convection coefficient of 400W/m²/K on the top outside edge and see that the average temperature is 31.4 °C.

In summary, we have found the convection boundary conditions that approximate the prescribed temperature boundary conditions used in part 1 of the HEATING PANEL analysis:

Convection boundary condition on hot side (inside):

Convection coefficient: 400W/m²/K; ambient temp. 85°C (358K);

This boundary condition produces an average temperature 80°C on the top inside edge.

Convection boundary condition on cold side (outside):

Convection coefficient: 100W/m²/K; ambient temp. 17°C (290K);

This boundary condition produces an average temperature 31.4 °C on the top outside edge.

The important difference is that in part 1, the prescribed temperature along the edges is uniform. Using convection boundary conditions, temperature along the edges is no longer uniform, which is more realistic.

Heat Power in the presence of convection boundary conditions is shown in (Figure 6-8).

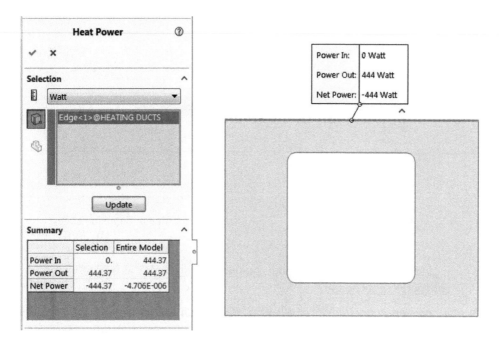

Figure 6-8: Heat power of one segment.

Each segment produces 444W heat power.

Compare this result to the result found in chapter 5.

Summary of studies completed

Model	Configuration	Study Name	Study Type
HEATING DUCTS.sldasm	*01 CAD*		
	02 Analysis Part 1		
	03 Analysis Part 2	*07 2D*	Thermal
		08 2D	Thermal

Figure 6-9: Names and types of studies completed in this chapter.

Notes:

7: Hot plate

Topics covered

- Transient thermal analysis
- Conductive heat transfer
- Convective heat transfer
- Heat power
- Thermostat
- Thermal inertia

Project description

A ceramic housing has an embedded heating element producing 1000W of thermal power. The power can be cycled on and off by a thermostat so the plate keeps the surface temperature at about 100°C. We will study temperature fluctuations with time for given thermostat settings.

Procedure

Open assembly model HOT PLATE in the *01 heating plate* configuration shown in Figure 7-1.

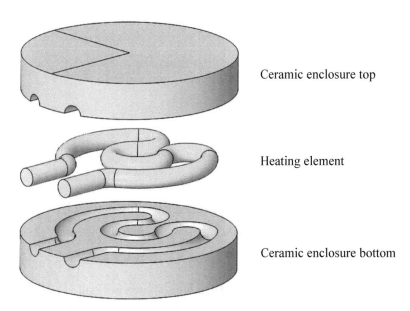

Ceramic enclosure top

Heating element

Ceramic enclosure bottom

Figure 7-1: HOT PLATE assembly in *01 heating plate* configuration.

The heating plate consists of two identical halves of ceramic enclosure that house a heating element. The split line on the enclosure face is there to position a thermostat.

Analysis of temperature fluctuations, while power is cycled on and off, requires transient thermal analysis. Create a Thermal study *01 transient* and set the study properties as shown in Figure 7-2.

Transient analysis —
Total time 2000s —
Time step 50s —

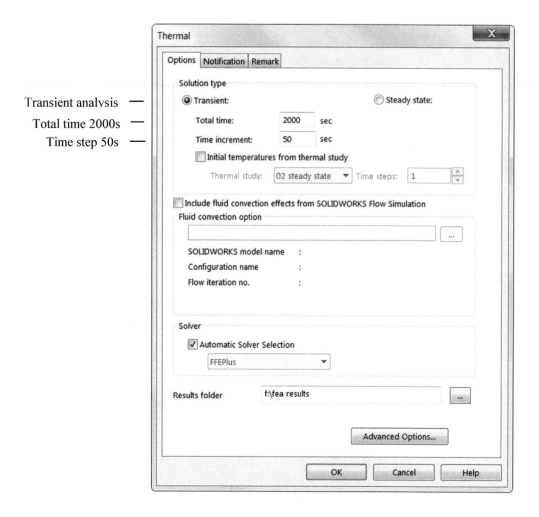

Figure 7-2: Properties of the transient thermal study.

The duration of analysis and time step usually requires some experimenting during trial runs.

Using the transient study settings shown in Figure 7-2, the analysis duration will be 2000s; it will be completed in 40 steps, 50 seconds each.

Define **Heat Power** and a **Thermostat** as shown in Figure 7-3.

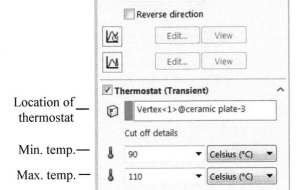
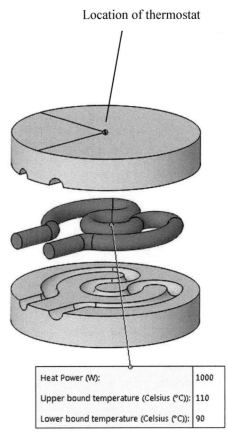

<u>Figure 7-3: Definition of a Heat Power controlled by a Thermostat.</u>

Select the heating element from the fly-out menu. Place the thermostat in the corner of the split line. The temperature in the location of the thermostat is the temperature that will control the power element on and off.

With the **Thermostat** setting set up as in Figure 7-3, power is turned on at time t = 0 and it stays on until the temperature reaches 110°C. At 110°C power is turned off until the temperature drops down to 90°C, then the power is turned on again. It is important to notice that this is the temperature on the outside of the heating plate, not the temperature of the heating element.

Define **Convection** as shown in Figure 7-4. Select all external faces of the ceramic housing. Do not select any face of the heating element.

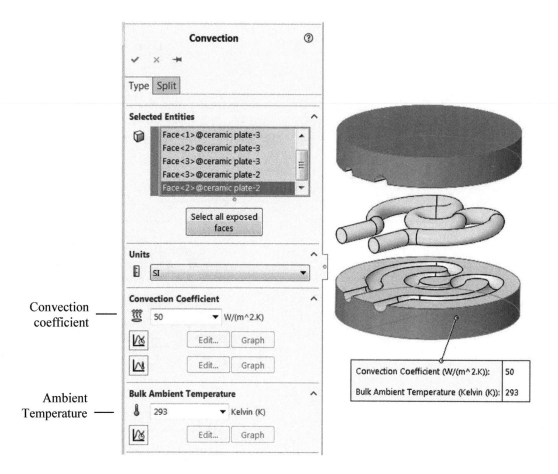

Figure 7-4: Definition of Convection.

The faces of the heating elements that stick out of the housing are insulated; do not define convection conditions on these faces.

Define an **Initial temperature** to all assembly components as shown in Figure 7-5.

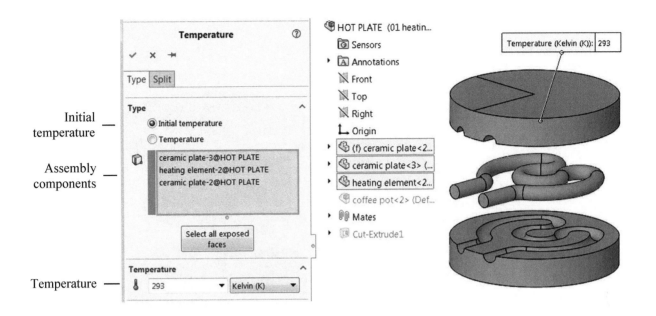

Figure 7-5: Definition of Initial temperature.

Select all three assembly components from the fly-out menu. Alternatively, you may select the assembly.

Notice that **Initial temperature** must be defined in **Transient** thermal analysis but not in **Steady State** thermal analysis.

Mesh the assembly model with a **Curvature based mesh** using settings shown in Figure 7-6.

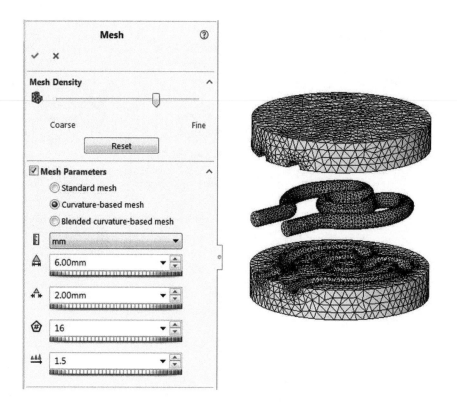

Figure 7-6: Curvature based mesh shown in exploded view.

The maximum element size is 6mm. A minimum of 16 elements in a circle gives an element turn angle of 22.5°.

A **Curvature based mesh** is more convenient to use because of the many curved faces present in the model. Review the mesh using exploded view to see how the heating element has been meshed. The mesh deficiencies would be easy to miss without the use of exploded view.

We assume an ideal bond between the heating element and ceramic plate; therefore, no thermal resistance is defined there.

The solution of study *01transient* completes in 50 time steps. If you want to reduce the solution time, use a less demanding mesh setting and/or switch to **Standard mesh**. You may also reduce the time of analysis. Once a solution completes, show the **Temperature** plot and follow the steps in Figure 7-7 to create a **Temperature Time History** graph.

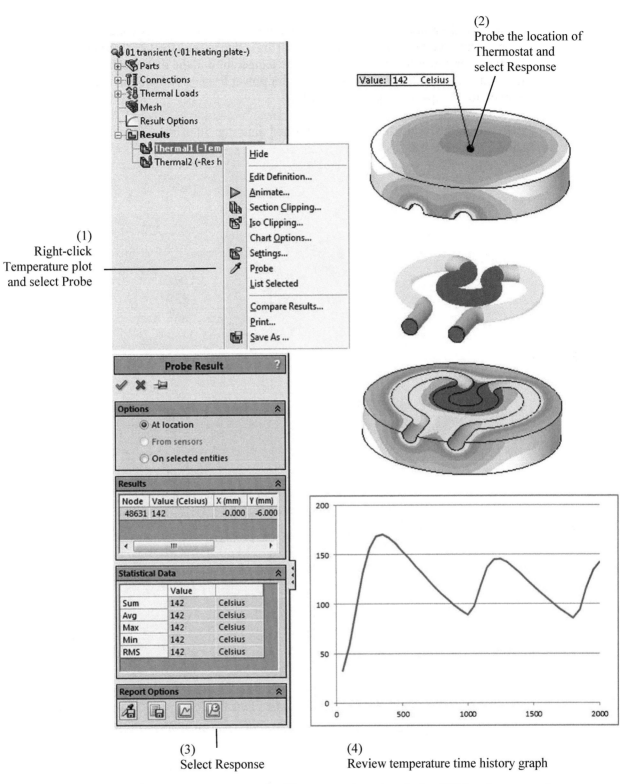

Figure 7-7: Temperature at the Thermostat location during 2000s after power has been first turned on.

After the initial warm-up period, the temperature fluctuates between 85C° and 145°C. Temperature time history graph has been formatted in Excel.

Notice that the range of temperature fluctuation of 85C° to 145°C is wider than what we have defined for our Thermostat. This is due to the effect of thermal inertia; after power has been turned off, the temperature in the thermostat location keeps on climbing for a while. After power has been turned on, the temperature keeps on dropping for a while.

A review of the results for the last performed time step 50 using exploded view is shown in Figure 7-8.

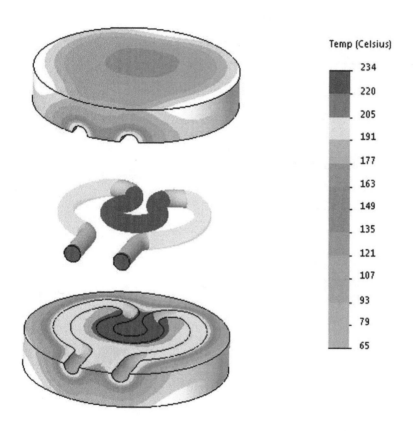

Figure 7-8: Temperature at the end of analysis, time step 50.

Notice the non-uniform temperature of the heating element.

Repeat the analysis with different **Heat Power** and **Thermostat** settings to see the effects of different settings on the temperature time history.

Due to mirror symmetry of the HOT PLATE assembly in configuration *01 heating plate* we could have conducted the same analysis on one half of the assembly model. We will take this approach in the continuation of this exercise, where we analyze the heating plate complete with a coffee pot.

Change configuration to *02 coffee pot* and notice that the assembly now includes the hot plate and a glass bowl filled with water. This assembly has mirror symmetry. An assembly cut reduces the model to one half (Figure 7-9).

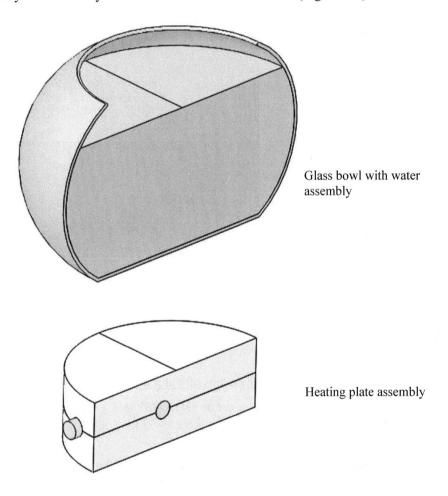

Glass bowl with water assembly

Heating plate assembly

Figure 7-9: HOT PLATE assembly in *01 coffee* configuration.

The model is shown in a partially exploded view.

Our objective is to find the steady state temperature of water when the glass bowl sits on top of the heating plate powered by a constant power of 200W. Create study *02 steady state*.

The most important part of this exercise is the definition of heat conductivity of water. Water has already defined properties in the CAD model (Figure 7-10), but these properties must be modified in **Simulation**. Why?

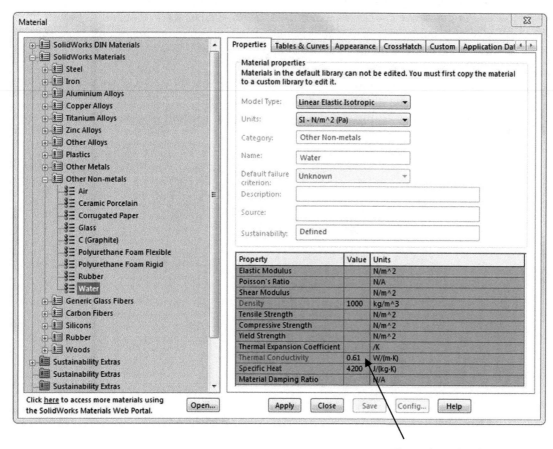

Thermal conductivity has to be changed from 0.61W/m/K) to 40W/m/K)

Figure 7-10: Thermal properties of water assigned to CAD model.

The thermal conductivity must be set to 40W/m/K prior to running analysis. To make this change follow steps in Figure 7-11.

The thermal conductivity as shown in Figure 7-10 does not take into consideration that water actually mixes while it is being heated, and convection, rather than conduction, is mainly responsible for heat transfer in water. Yet, the only mechanism of heat transfer that can be modeled in a solid body (and this is how water is modeled here) is conduction. To account for convection inside the volume of water, which we cannot model, we have to define an artificially high thermal conduction of 40W/m/K. This value has been found by testing. Modification of thermal conductivity requires definition of a custom material. Follow the steps shown in Figure 7-11. Make sure you do the modification in the **Simulation** study, not in the CAD assembly model.

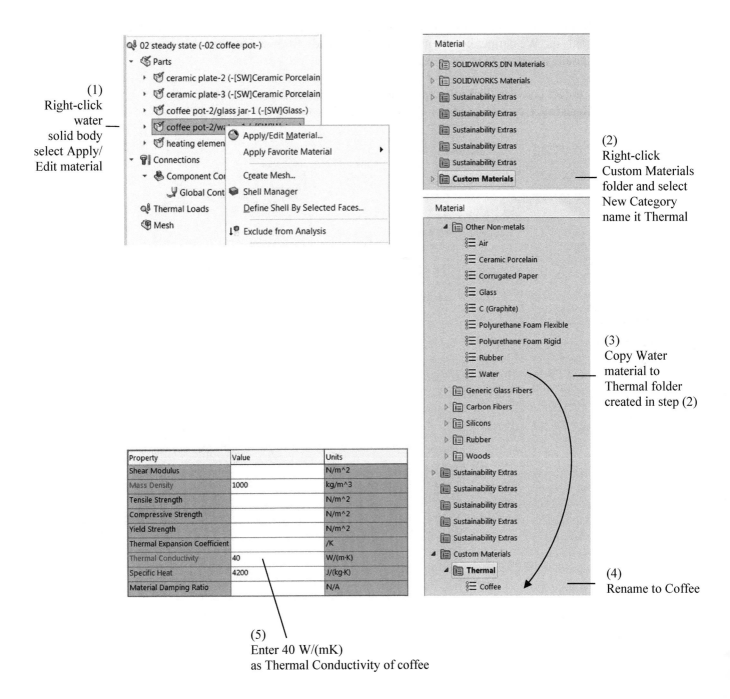

(1)
Right-click
water
solid body
select Apply/
Edit material

(2)
Right-click
Custom Materials
folder and select
New Category
name it Thermal

(3)
Copy Water
material to
Thermal folder
created in step (2)

(4)
Rename to Coffee

(5)
Enter 40 W/(mK)
as Thermal Conductivity of coffee

Figure 7-11: Definition of custom material properties of water in the Simulation study.

Name the modified material coffee. Define Thermal Conductivity 40W/m/K; keep mass density 1000kg/m³ and specific heat 4200J/kg.

Notice that mass density and specific heat are not required in steady state thermal analysis. We define them to make the model ready for transient thermal analysis should you wish to run it later.

Contact between the heating plate and glass is not perfect and a **Thermal Resistance** contact condition must be defined as shown in Figure 7-12.

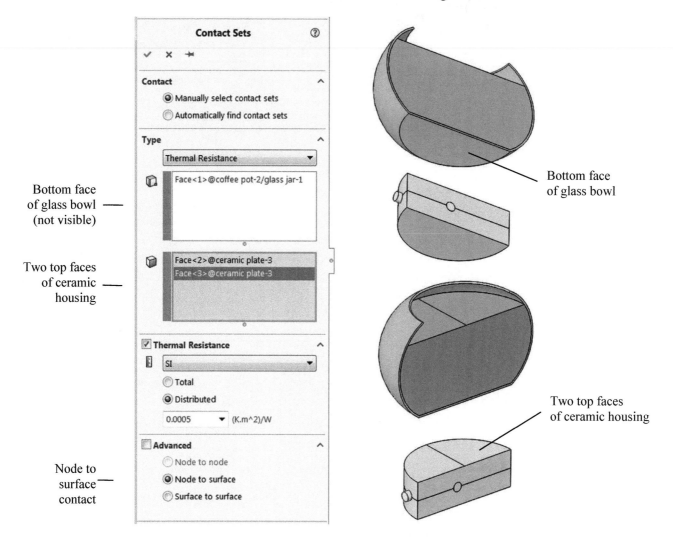

Figure 7-12: Definition of Thermal Resistance between the heating plate and the glass bowl.

The value of Thermal Resistance can be determined in testing by measuring the temperature difference between the faces that remain in contact.

In the steady state analysis, we assume that the heating element produces 200W of power. One half of the heating element is modeled; therefore, define 100W. To apply **Heat Power**, refer to Figure 7-3. Since this is a **Steady State** analysis, there is no need to define a **Thermostat**.

Apply **Convection** as shown in Figure 7-13.

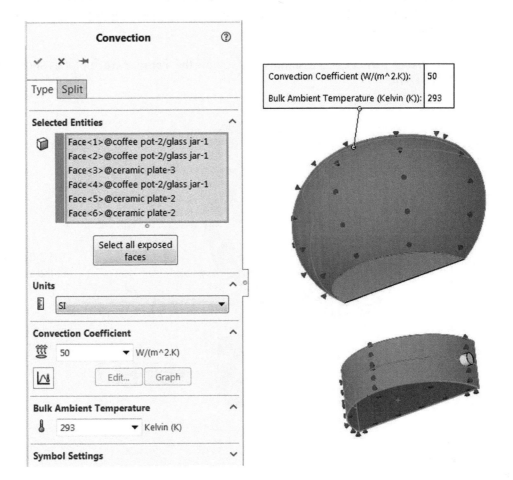

Figure 7-13: Definition of Convection.

Define a convection boundary condition for all faces exposed to air including water and excluding faces of the heating elements that stick out of the housing. Do not apply convection boundary conditions to any face in the plane of symmetry; that would destroy the symmetry.

Due to symmetry, there is no heat flow in the direction normal to the plane of symmetry. The symmetry boundary condition is enforced by not defining convection on the faces in the plane of symmetry.

We have to point out that while defining **Convection** in **SOLIDWORKS Simulation** is easily done, finding the actual value is often the most difficult part of thermal analysis. Applying the same $50 W/(m^2 K)$ to all faces including water is a modeling simplification which has to be supported by testing.

Having defined **Heat Power** and **Convection**, we have already defined the mechanisms of heat transfer. No Initial **Temperature** is required in **Steady State** thermal analysis. Mesh the model using mesh settings shown in Figure 7-6 and run the solution.

Hide *Water* part in CAD assembly and review the **Temperature** plot as shown in Figure 7-14.

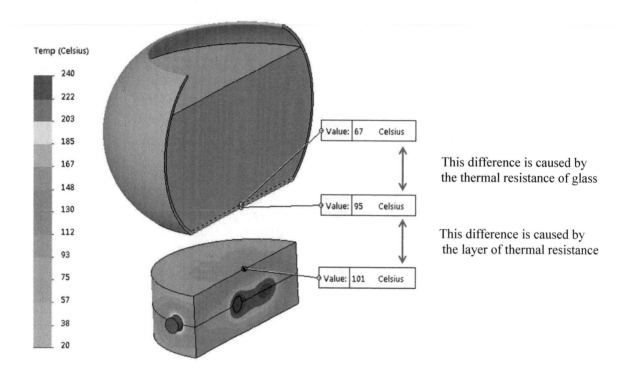

Figure 7-14: Temperature results; probing shows the effect of the layer of thermal resistance and thermal resistance of glass.

Notice that the probing location on the side of the hot plate is well defined by split line, but on the side of the coffee pot you have to "eyeball" probing location. Therefore, your probed result may be slightly different from these shown above.

When probing any results, you may deselect Show Node/Element Number and Show X, Y, Z Location for an uncluttered display.

The unit of thermal resistance is $K/m^2/W$; notice that is the inverse of the unit of thermal conductivity $W/m^2/K$.

Summary of studies completed

Model	Configuration	Study Name	Study Type
HOT PLATE.sldasm	*01 heating plate*	*01 transient*	Thermal
	02 coffee pot	*02 steady state*	Thermal
	03 full model		

Figure 7-15: Names and types of studies completed in this chapter.

Notes:

8: Thermal and thermal stress analysis of a coffee mug

Topics covered

- ❑ Transient thermal analysis
- ❑ Thermal stress analysis
- ❑ Thermal symmetry boundary conditions
- ❑ Structural symmetry boundary conditions
- ❑ Use of soft springs

Project description

A ceramic coffee mug at room temperature is filled with boiling water for long enough to reach a steady state temperature, then the water is poured out and the mug is left to cool down. We wish to find thermal stresses in the mug as a function of time during the cooling down process.

Procedure

Review MUG model (Figure 8-1).

Configuration *01 full*   Configuration *02 half*

Figure 8-1: Coffee mug in two configurations.

Configuration 02 half will be used for analysis.

First, we conduct a **Steady state** thermal analysis to find the initial conditions for the cool down process. Taking advantage of symmetry, we will use the *02 half* configuration. Switch to *02 half* configuration and create a **Thermal** study called *01 steady state*. Due to a high convection coefficient between water and porcelain, the inside of the mug will reach almost 100°C, the temperature of boiling water. This can be modeled as a prescribed temperature (Figure 8-2).

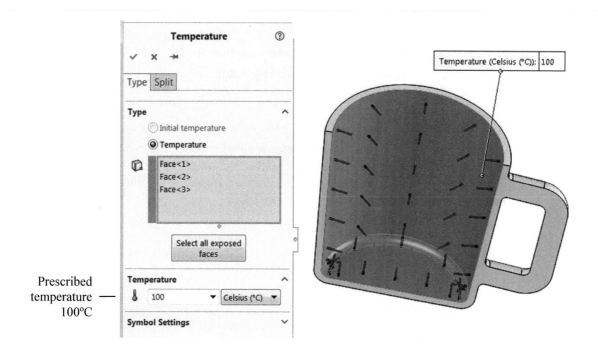

Prescribed temperature 100°C

Figure 8-2: Prescribed temperature boundary conditions on inside faces of the mug.

A prescribed temperature is defined on the three inside faces. No prescribed temperature is defined on the faces in the plane of symmetry.

Heat enters the mug through faces where prescribed temperature is defined, travels by conduction through the porcelain, and exits to the surrounding air by convection. Define a convection coefficient on the outside faces including the top face as shown in Figure 8-3.

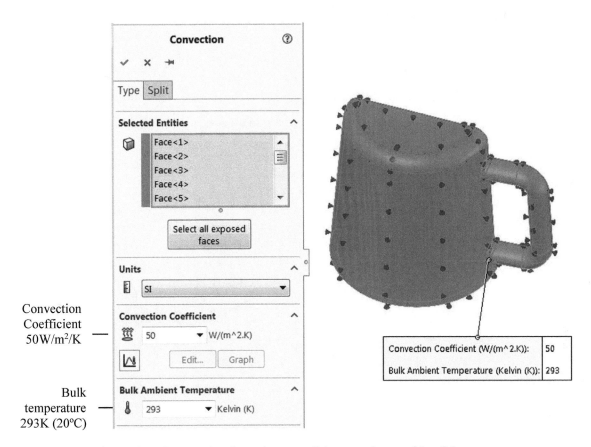

Convection Coefficient 50W/m²/K

Bulk temperature 293K (20°C)

Figure 8-3: Convection boundary conditions on the outside of the mug.

No convection boundary conditions are defined on the faces in the plane of symmetry.

By not defining any convection conditions on the faces in the plane of symmetry, we enforce thermal symmetry boundary conditions, which requires that no heat flows in the direction normal to the plane of symmetry.

Before meshing the model, we must remember that in the following thermal analysis, we will conduct the analysis of thermal stresses. In thermal analysis we will study temperature results, which usually does not require "too much" mesh refinement. It is analogous to the analysis of displacements only (and not stresses) in structural analysis. However, in thermal stress analysis, which is a structural analysis, we will study stresses. The mesh must be identical in thermal and thermal stress analyses and therefore, it must satisfy the more demanding requirements of thermal stress analysis. Define mesh controls (as shown in Figure 8-4) on rounds where stress concentrations are expected.

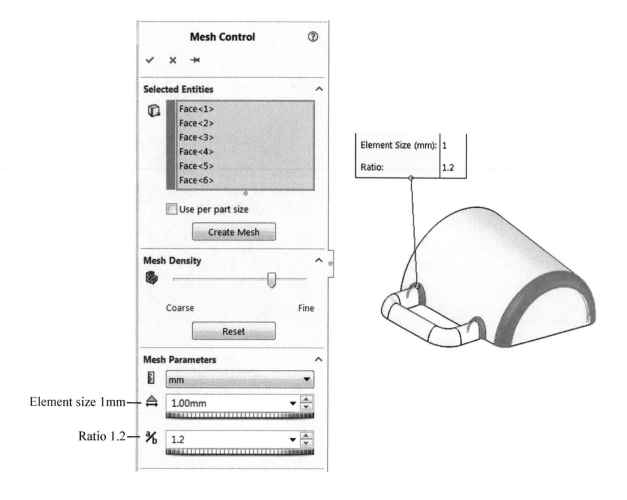

Figure 8-4: Mesh controls are defined on fillets where the ear connects to cup and at the bottom of mug on the outside and the inside round face which is invisible in this illustration.

1mm element size ensures low turn angle; a 1.2 transition ratio a/b ensures a smooth transition between locally refined mesh and the rest of the model.

Use a global element size of 1.8mm to create a mesh with two elements across the wall thickness. Run the solution of study *01 steady state* and review **Temperature** results (Figure 8-5).

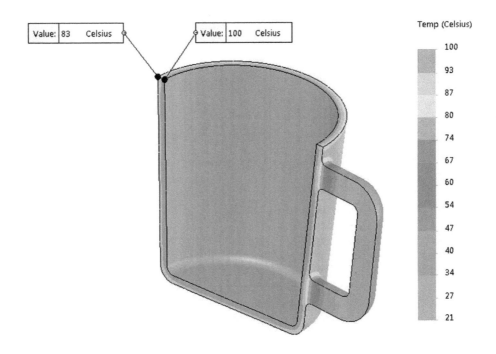

Figure 8-5: Steady state temperature results; notice a low temperature gradient across the mug wall as shown in the probed locations.

User defined colors have been used in this plot to improve the appearance of this black and white illustration. This technique is used in many fringe plots presented in this book.

The temperature results from the steady state thermal analysis conducted in study *01 steady state* are the initial conditions for transient thermal analysis.

Copy study *01 steady state* into *02 transient*. To change from steady state to transient thermal analysis, edit the study properties as shown in Figure 8-6.

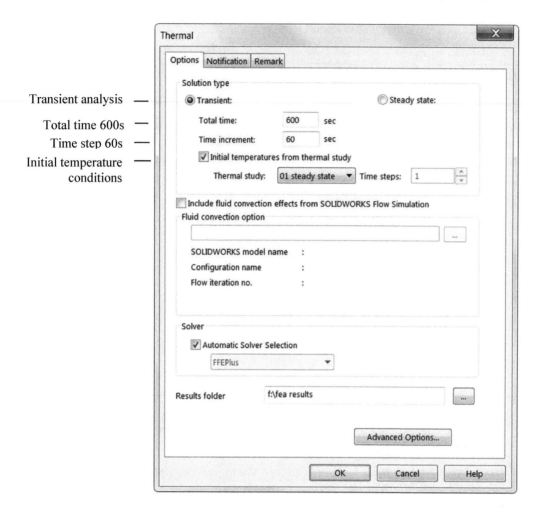

Transient analysis —

Total time 600s —

Time step 60s —

Initial temperature —
conditions

Figure 8-6: Properties of *02 transient* study. The definition includes the duration and time step. Initial temperature conditions are taken from the previously completed steady state thermal study *01 steady state*.

A total time of 600s and a time step of 60s mean that the analysis will be completed in 10 steps.

Delete the prescribed temperature (water has been poured out) and edit convection conditions to include faces where prescribed temperature was defined before (the inside faces are now exposed to air).

Run the transient solution and display a **Temperature** plot for step 4 or other step(s) of your choice (Figure 8-7).

Figure 8-7: Temperature results from time step 4.

Review other steps to see the change of temperature distribution over time.

Probe the temperature plot from any time step to produce temperature time history as shown in Figure 8-8.

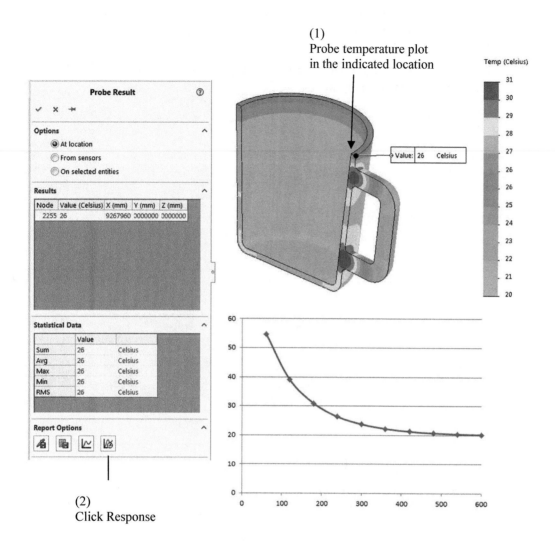

Figure 8-8: Temperature time history from all performed steps 1 through 10.

It can be seen that after 600s the mug cools down almost reaching ambient temperature. The graph has been formatted in Excel.

With temperature results from all 10 steps, we may now find out how thermal stresses change over time. Create a static study *03 stress step 1* and define study properties as shown in Figure 8-9.

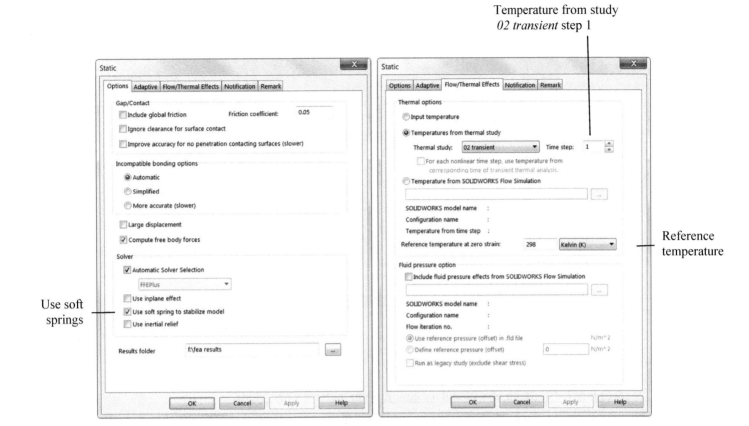

Options   Flow/Thermal Effects

Figure 8-9: Properties of a static study used for analysis of thermal stresses.

Select Use soft springs to stabilize the model. The reference temperature at zero strain equals the ambient temperature used in thermal studies.

While thermal boundary condition requires no action (nothing has to be defined on the faces of symmetry), structural boundary conditions must be defined to eliminate out-of-plane motions. Follow Figure 8-10 to define structural symmetry boundary conditions.

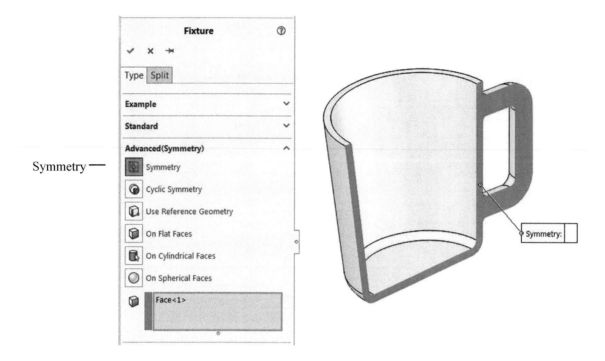

Figure 8-10: Symmetry boundary condition in thermal stress study.

Symmetry boundary conditions suppress translational degrees of freedom of nodes located on the plane of symmetry in the direction normal to the plane.

Symmetry boundary conditions will not restrain the model completely. Under the restraints shown in Figure 8-10, the model is able to slide in two directions along the plane of symmetry and to rotate about the axis perpendicular to the plane of symmetry, meaning that the model has 3 rigid body motions (RBMs). We could easily add a **Fixed** restraint somewhere to the model but that would change stress results; remember that we want to see thermal stresses which are caused by uneven temperature distribution throughout the model. This is why we use the soft springs option (Figure 8-9) to eliminate these RBMs.

In a thermal stress analysis, nodal temperatures are imported from the thermal to structural (here static) study and that requires that the meshes in the thermal and structural studies are identical. The best way to make sure the meshes are identical is to copy the mesh from study *02 transient* to study *03 stress step 1*.

Run study *03 stress step 1* and display the stress results. Remember that von Mises stress should not be used for evaluating stress results when brittle material (ceramic porcelain) is analyzed; instead, use the maximum principal stress. Refer to chapter 1 of "**Engineering Analysis with SOLIDWORKS Simulation**" for more information. Repeat the analysis of results using next few time steps to see how stresses change while the mug is cooling down.

P1 stress results from step 1 are shown in Figure 8-11.

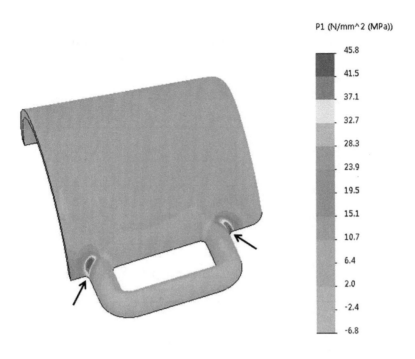

Figure 8-11: P1 stress results from step 1.

Stress concentration locations are indicated by arrows.

Display stress results with overlaid mesh to see that the mesh is at best marginal to model highly localized stress concentrations. You may want to repeat this exercise with a more refined mesh.

Review the stress results from other steps to notice that the highest stress of 41MPa is at the beginning of the cooling process. To calculate the factor of safety, the maximum P1 stress should be compared to the ultimate strength of the ceramic porcelain (173MPa). This calculation gives a factor of safety of 4.2.

To complete this exercise, animate displacement results from step 1 to examine the pattern of thermal deformation. Next, construct a graph showing P1 stress as function of time (steps 1 through 10); use location indicated by sensor defines in **SOLIDWORKS** model.

Notice that deformations and stresses are caused by thermal effects only. Structural load or restraints have not been used in this analysis.

Summary of studies completed

Model	Configuration	Study Name	Study Type
MUG.sldprt	01 full		
	02 half	01 steady state	Thermal
		02 transient	Thermal
		03 stress step 1	Static

Figure 8-12: Names and types of studies completed in this chapter.

9: Thermal buckling analysis of a link

Topics covered

- Buckling caused by thermal effects
- Interpretation of Buckling Load Factor

Project description

A rectangular steel link is restrained by hinges on both ends. Its temperature is raised from the initial temperature of 298K to the final temperature 358K in six 10K steps. We want to know how the Buckling Load Factor (BLF) changes with the change of temperature.

Procedure

The analysis will be conducted on part model LINK TH. The temperature of link at every step is uniform and that can be defined in buckling analysis. There is no need for a pre-requisite thermal analysis. If the temperature was distributed non-uniformly, we would have to find temperature distribution in a prerequisite thermal analysis.

Create a *Buckling* study called *b308* and verify that the study properties indicate 298K as the reference temperature at zero strain (Figure 9-1). Thermal options default to **Input temperature** so no action is required.

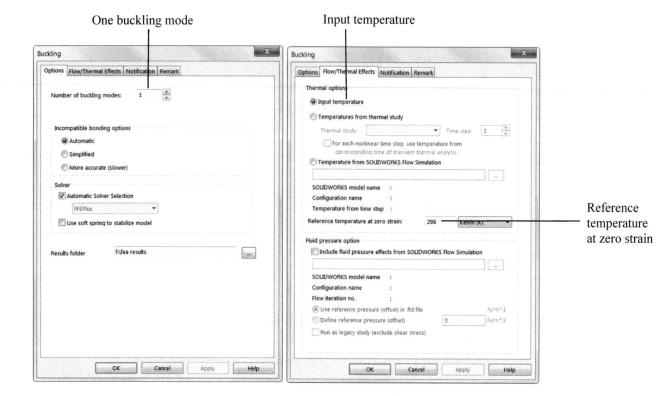

Figure 9-1: Properties of buckling study.

Reference temperature at zero strain is 298K. Selections shown in this illustration are all default.

Define restraints as shown in Figure 9-2.

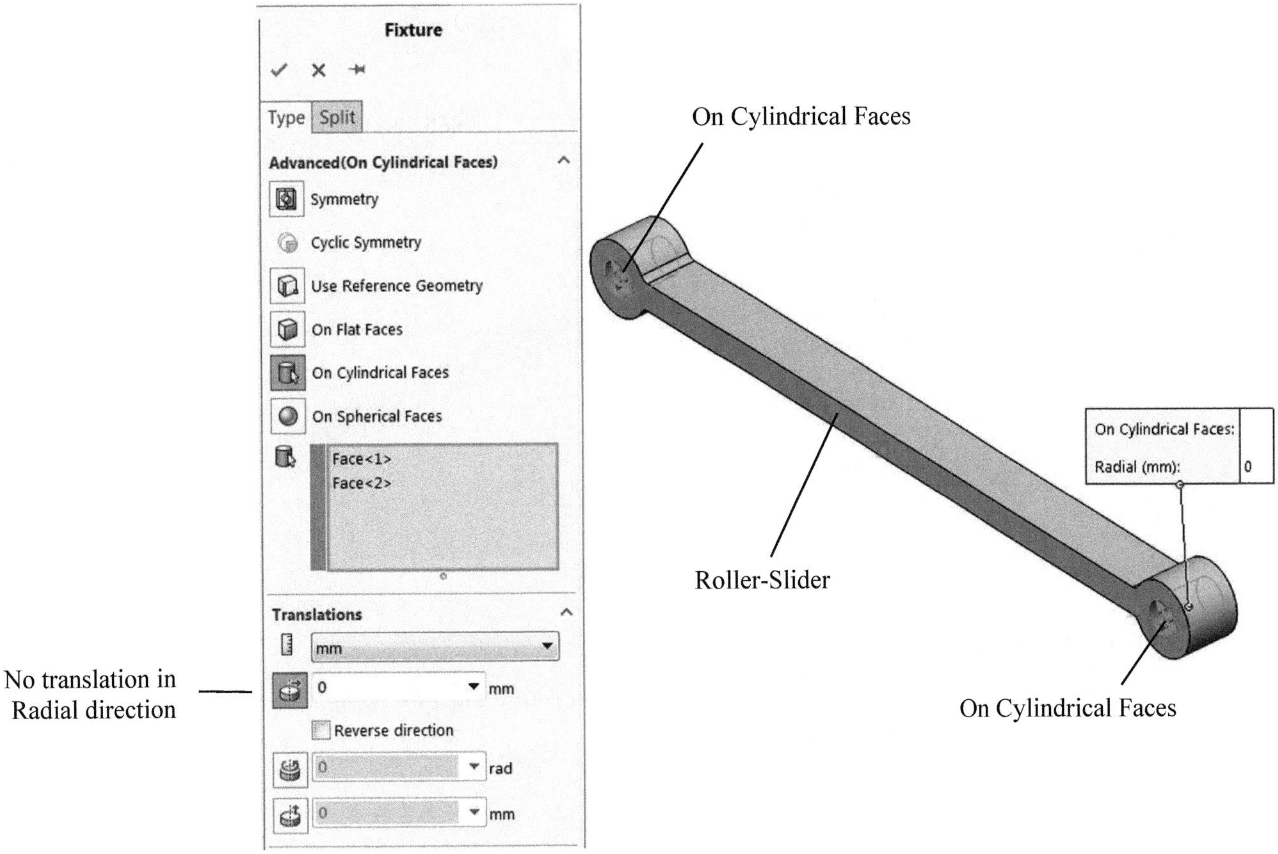

<u>Figure 9-2: Link with hinge supports on both ends.</u>

Hinge supports are modeled as On Cylindrical Faces with eliminated translation in the Radial direction. Roller-Slider restraint (definition not shown in this illustration) is applied to one side face to restrain translation in the axial direction of hinges.

Apply **Temperature** of 308K to the solid body as shown in Figure 9-3.

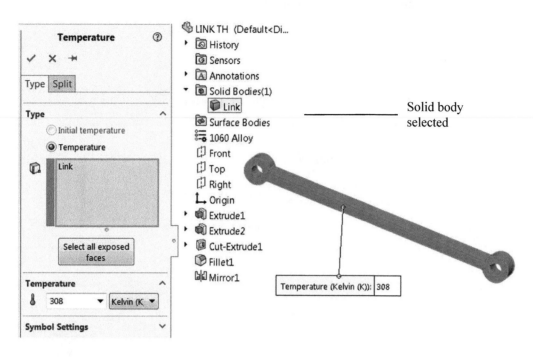

Figure 9-3: Temperature definition.

Select body called link from the fly-out menu. Temperature must be applied to the solid body rather than to faces.

Run *Buckling* study *b308* using default element size and display the first buckling mode (Figure 9-4).

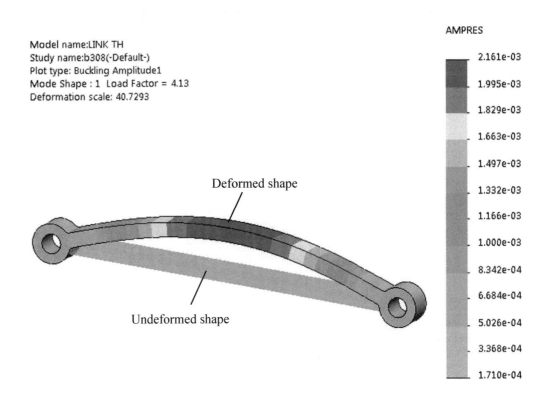

Model name:LINK TH
Study name:b308(-Default-)
Plot type: Buckling Amplitude1
Mode Shape : 1 Load Factor = 4.13
Deformation scale: 40.7293

AMPRES

2.161e-03
1.995e-03
1.829e-03
1.663e-03
1.497e-03
1.332e-03
1.166e-03
1.000e-03
8.342e-04
6.684e-04
5.026e-04
3.368e-04
1.710e-04

Deformed shape

Undeformed shape

Figure 9-4: The first buckling shape, buckling load factor BLF=4.13.

Undeformed shape is shown; the numerical values indicate the resultant amplitude; these values can be used only to find ratio between displacement magnitudes in different locations; absolute numbers are meaningless. Animate results to see that the deformed shape can be either "up" or "down". A linear buckling analysis cannot determine which way the model buckles.

The first buckling mode shown in Figure 9-4 has a BLF of 4.13. The BLF is a ratio between the load causing buckling and the applied load:

$$BLF = \frac{Buckling\ load}{Applied\ load} = 4.13$$

The only loads on the model are the hinge reactions caused by the fact that hinges prevent thermal expansion of the link. BLF=4.13 means that the reaction force would have to increase by a factor of 4.13 in order for buckling to take place. It does not refer directly to temperature.

To find those reactions, we need to run a *Static* study. Create static study *s308*; copy restraints, loads (**Temperature**), and mesh from the completed buckling study *b308* and obtain a static solution. Right-click the *Results* folder, select **List result force** and review the reaction forces in hinges (Figure 9-5).

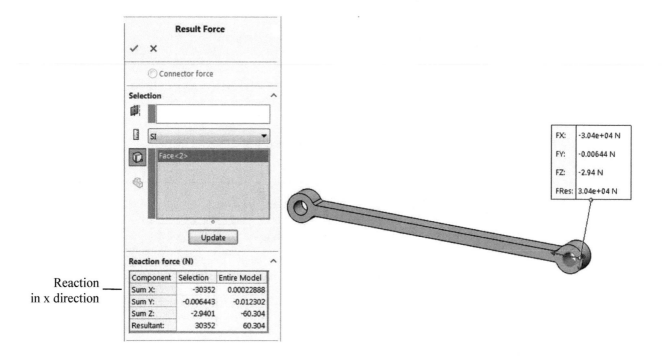

Figure 9-5: Reaction at one end.

Reactions in directions y and z are caused by discretization errors. Select the other hinge to verify that reactions in both hinges are opposite and equal.

Copy study *b308* and *s308* into *b318* and *s318* and make the link temperature equal to 318K in these studies. Read the BLF from the buckling study and reaction force from the static study. Repeat the analysis in 10K increments creating studies *b328* and *s328*, *b338* and *s338*, *b348* and *s348*, *b358* and *s358*. Results are summarized in Figure 9-6.

Temperature K	BLF	Reaction force N
308	4.13	30352
318	2.07	60704
328	1.38	91057
338	1.03	121410
348	0.83	151760
358	0.69	182110

Figure 9-6: Summary of results of buckling and static studies.

BLF results come from buckling studies, reaction forces come from static studies.

As shown in Figure 9-6, the BLF for temperature of 338K is approximately equal to 1. It means that the increase of temperature of the link from initial 298K to 338K will cause buckling. The reaction force for temperature 338K is 121410N.

Figure 9-7 shows the BLF as a function of link temperature.

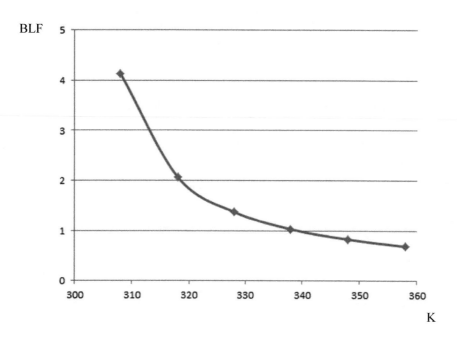

Figure 9-7: Buckling Load Factor as a function of link temperature.

Summary of studies completed

Model	Configuration	Study Name	Study Type
LINK TH.sldprt	*Default*	*b308*	Buckling
		b318	Buckling
		b328	Buckling
		b338	Buckling
		b348	Buckling
		b358	Buckling
		s308	Static
		s318	Static
		s328	Static
		s338	Static
		s348	Static
		s358	Static

Figure 9-8: Names and types of studies completed in this chapter.

Notes:

10: Thermal analysis of a heat sink

Topics covered

- ❏ Analysis of an assembly
- ❏ Thermal contact conditions
- ❏ Steady state thermal analysis
- ❏ Transient thermal analysis
- ❏ Thermal resistance layer
- ❏ Thermal symmetry boundary conditions

Project description

We need to find the relationship between the height of cooling fins of an aluminum radiator used to cool an electronic component and the temperature of that component. We will demonstrate that fins that are too short do not provide adequate cooling while fins that are too long do not offer additional advantages in cooling.

Three configurations of the HEAT SINK TH assembly model are shown in Figure 10-1.

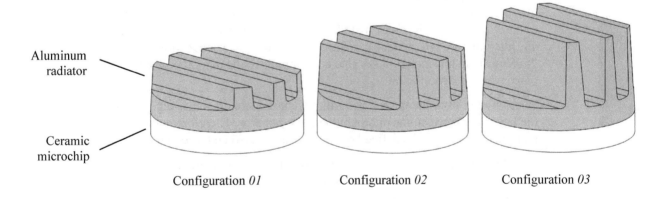

Aluminum radiator

Ceramic microchip

Configuration *01*   Configuration *02*   Configuration *03*

Figure 10-1: Assembly model of a microchip with a radiator.

HEAT SINK TH assembly consists of two components: a ceramic microchip and an aluminum radiator. All heat generated by the microchip dissipates through the radiator. The two faces of the microchip, the cylindrical side face and the bottom flat face (not visible in this illustration), are insulated. The difference between the three configurations is the height of the cooling fins.

The double symmetry of the HEAT SINK TH model offers an opportunity to simplify the analysis by using one half or one quarter of the model; here we decide to use one half of the model by working with configurations shown in Figure 10-2.

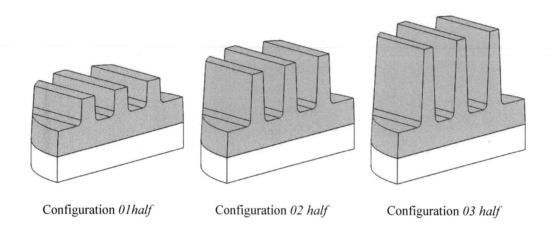

Configuration *01half*   Configuration *02 half*   Configuration *03 half*

Figure 10-2: *01 half, 02 half* and *03 half* configurations used for analysis.

We use half of the model to practice thermal symmetry boundary conditions.

The ceramic microchip generates a heat power of 50W and the aluminum radiator dissipates this heat. The ambient temperature is 27°C (300K). Heat is dissipated to the environment by convection through all exposed faces of the radiator. The microchip is insulated; it cannot dissipate heat directly to ambient air, but only through the face touching the radiator. The convection coefficient is 25 W/m²/K in this model; this value corresponds to the natural convection taking place without a cooling fan.

Heat flowing from the microchip to the radiator encounters thermal resistance on the boundary between the microchip and the radiator. Therefore, a thermal resistance layer must be defined on the interface between these two components.

To select the height of the cooling fins, we will determine the temperature and heat flux of the assembly in steady state conditions for the model in three configurations. This will require steady state thermal analysis.

Once we select the height of the cooling fins, we will study the temperature in the assembly as a function of time when the assembly is initially at room temperature and the power is cycled on and off every 10s. This will require transient thermal analysis.

Procedure

Activate configuration *01 half* and create a *Thermal* study called *01 half*. Notice that the *Solids* folder contains two bodies corresponding to the two assembly components with material properties already assigned: **Ceramic Porcelain** to the microchip and **1060Alloy** to the radiator.

Before proceeding, we need to look at the *Connections* folder. By default, the **Global Contact** between parts in an assembly is **Bonded**. As the name implies, all parts in the assembly behave as one and heat encounters no resistance flowing from one part to another. We need to override the **Global Contact** condition by defining a **Contact Set**.

Right-click the *Connections* folder and select **Contact Set** to open the **Contact Set** window shown in Figure 10-3. Select **Thermal Resistance** with **Node to surface** as the contact type. This contact condition overrides the global **Bonded** contact condition. Use the exploded view to select the contacting faces and enter a **Distributed Thermal Resistance** of $0.001 Km^2/W$. The value of thermal resistance may be obtained by testing.

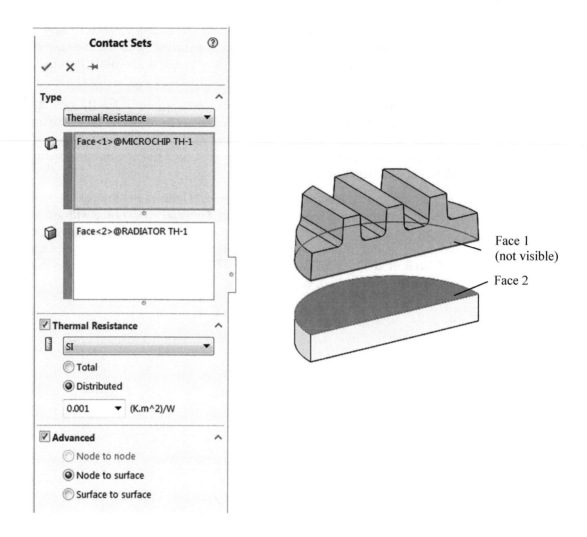

Figure 10-3: Definition of Thermal Resistance Contact Set.

We need to define a local Contact Set for contacting faces in order to introduce a thermal resistance layer between the faces. Use the exploded model view to define the Contact Set. This can only be done as a local Contact Set using the Node to surface option in Advanced options.

Next, specify the heat power generated in the microchip. To do this, right-click the *Thermal loads* folder to open the pop-up menu. Select **Heat Power...** to open the **Heat Power** window (Figure 10-4), and then from the **SOLIDWORKS** fly-out menu select the *MICROCHIP TH* assembly component and define a 25W heat power which is one half of the heat power generated in the microchip. This applies heat power to the entire volume of the selected component.

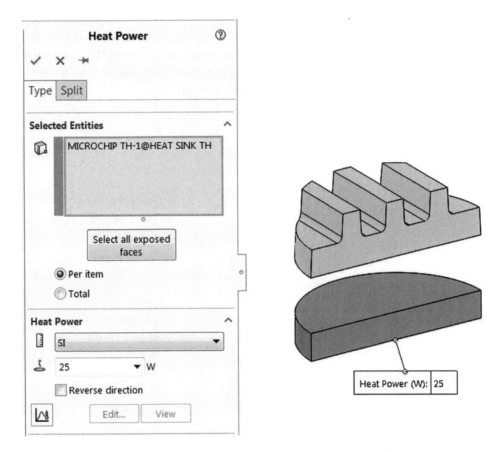

Figure 10-4: Heat Power is applied to the entire volume of the microchip.

Notice that reversing the direction in the Heat Power definition window would produce cooling.

So far, we have assigned material properties to each component, a heat source, and a thermal resistance layer. In order for heat to flow, we must also establish a mechanism for heat to escape the model. This is accomplished by defining convective boundary conditions.

Right-click the *Thermal Loads* folder to open a pop-up menu and select **Convection...** to open the **Convection** window (Figure 10-5). Select all faces of the RADIATOR TH except the one touching the MICROCHIP TH and the face representing the symmetry plane. Do not select any face of the MICROCHIP TH. Enter $25 W/m^2/K$ as the value of the convection coefficient for all selected faces and enter the **Bulk Ambient Temperature** as 300K.

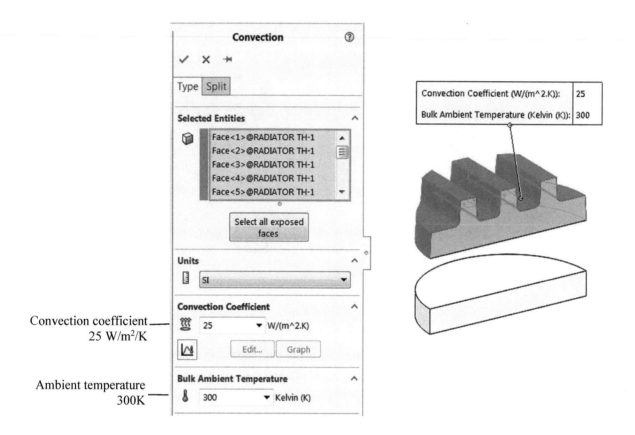

Figure 10-5: Definition of convective boundary conditions on the outside faces of the radiator.

Use this window to specify both the Convection Coefficient and the Ambient Temperature. Do not define convection on the face in the plane of symmetry, the face in contact with the microchip, or to any faces of the microchip.

The last step before solving is creating the mesh. For accurate heat flux results, we need several elements modeling the fillet curvature. This can be done either using a **Standard Mesh** with mesh controls applied to all six fillets or using a **Curvature based mesh** as shown in Figure 10-6; we will use this method.

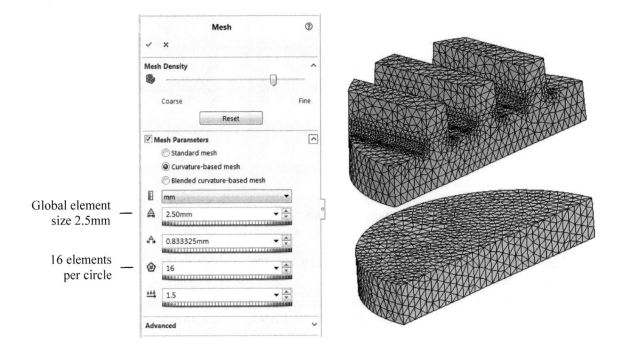

Global element size 2.5mm

16 elements per circle

Figure 10-6: A curvature based mesh assures correct meshing of fillets without the use of mesh controls. Use 16 elements per circle to ensure a low turn angle.

Notice that a curvature based mesh produces a low element turn angle at the base of each fin.

We use a **Curvature based mesh** with settings shown in Figure 10-6 for all **Steady State** studies, even though the heat flux will not be analyzed. The **Transient** analysis will be performed with a default standard mesh.

Once the solution is ready, examine the temperature results. Repeat the analysis for *02 half* and *03 half* configurations; the summary is shown in Figure 10-7.

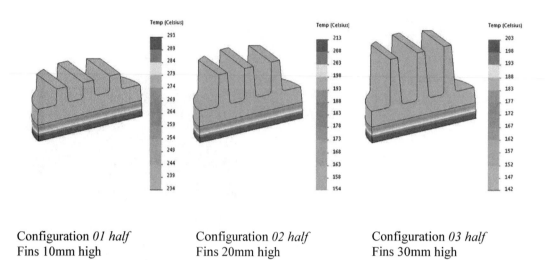

Configuration *01 half*
Fins 10mm high

Configuration *02 half*
Fins 20mm high

Configuration *03 half*
Fins 30mm high

Figure 10-7: The temperature plots from the steady state analyses in three configurations.

Notice that the temperature of the radiator is practically uniform due to high conductivity of aluminum. Temperature in the microchip is the highest at the bottom.

To decide if the fin height should be increased past 30mm, we present the results in a graph, shown in Figure 10-8.

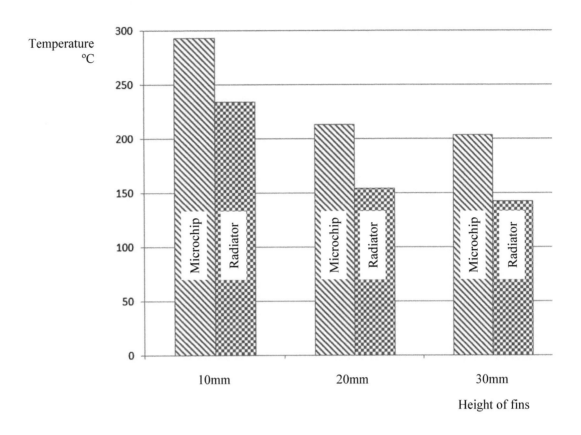

Figure 10-8: The summary of the maximum temperature results from the steady state analyses in three configurations.

Notice that the increase in fin length from 10mm to 20mm has a much stronger effect on the temperature of the microchip and the radiator than the increase from 20mm to 30mm.

A review of the graph shown in Figure 10-8 proves that increasing the height of fins follows the law of diminishing returns. Here, we decide that 30mm high fins offer sufficient cooling and proceed with a transient analysis of the model in the *03 half* configuration.

Copy the study *03 half* into *03 half transient* and make it into a **Transient** study by defining properties as shown in Figure 10-9.

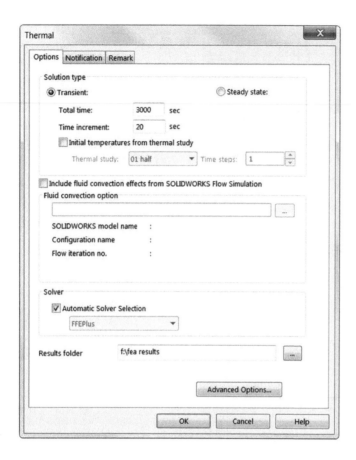

Figure 10-9: Properties of the transient thermal study.

Analysis will be conducted for 3000s in 20s intervals.

Overwrite the mesh copied from study *03 half* by meshing the assembly with a standard mesh using a default global element size to reduce the number of elements. Large elements are used to reduce solution time in this analysis and are acceptable if we limit the analysis of results to temperatures. However, we would not be able to rely on heat flux results.

As in any transient thermal analysis, the **Initial temperature** must be defined for all components. Define it as shown in Figure 10-10.

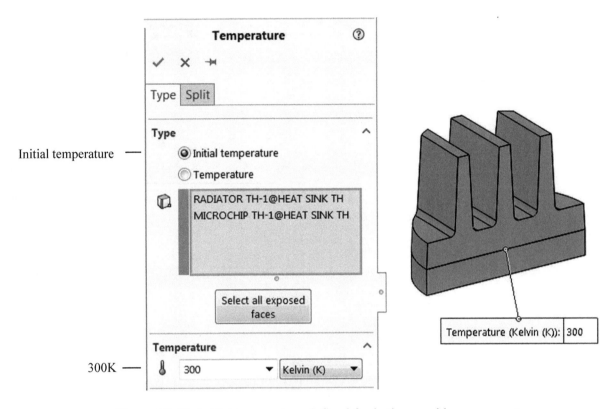

Figure 10-10: Initial temperature defined for both assembly components.

Select both components from the fly-out menu not shown in this illustration.

The definition of the convection coefficient remains unchanged but the heat power definition must be modified to make it a function of time.

In this analysis power cycled on and off every 100s.

Open the Excel file located in the same folder as HEAT SINK TH and follow the steps shown in Figure 10-11.

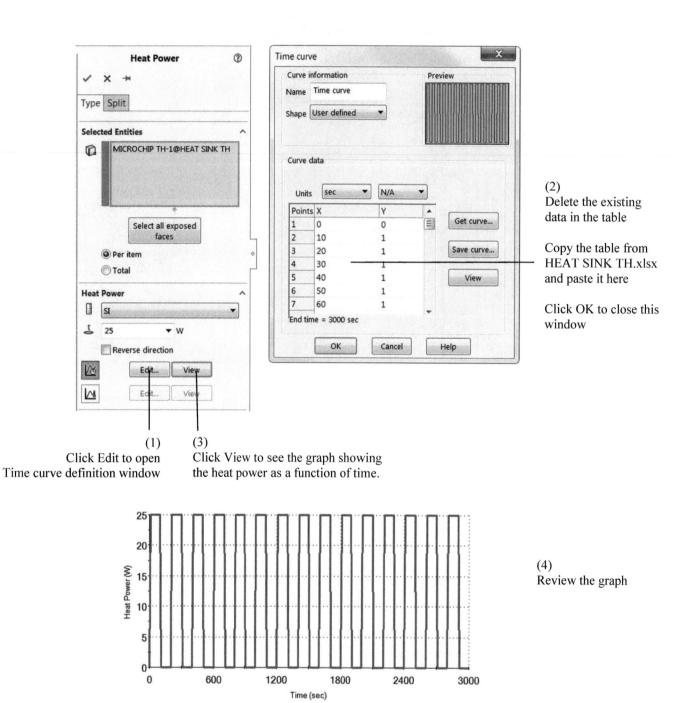

(2)
Delete the existing data in the table

Copy the table from HEAT SINK TH.xlsx and paste it here

Click OK to close this window

(1)
Click Edit to open Time curve definition window

(3)
Click View to see the graph showing the heat power as a function of time.

(4)
Review the graph

Figure 10-11: Definition of transient heat power.

Copy the table from HEAT SINK TH.xlsx.

Run the analysis and display a temperature plot from any time step; it does not have to be the last step. Probe the radiator and create a plot of temperature as a function of time by following the steps shown in Figure 10-12.

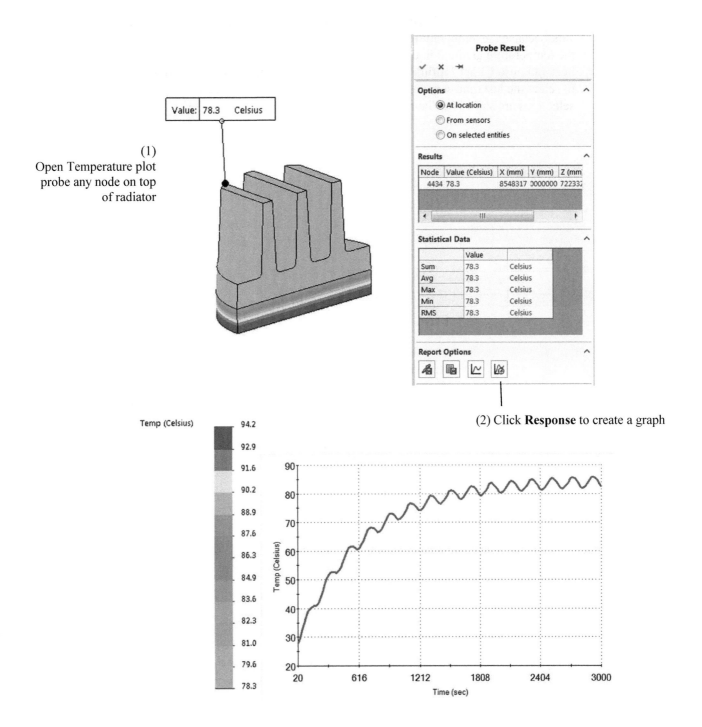

Figure 10-12: Temperature of radiator as function of time.

The temperature of radiator is almost uniform; therefore, any point away from the microchip may be selected for probing temperature results.

The graph in Figure 10-12 shows that after 3000s, the temperature fluctuations stabilize in the range of 82-86°C.

All temperature plots defined so far have shown uniform temperature in the heat sink because the temperature gradient in the heat sink is very low as compared to the temperature gradient in the microchip. To see the temperature gradient in the heat sink, use **Chart Options** to adjust the range of temperature. Use **Definition** to select the last time step and units; use **Settings** to select discrete fringes. All selections are shown in Figure 10-13.

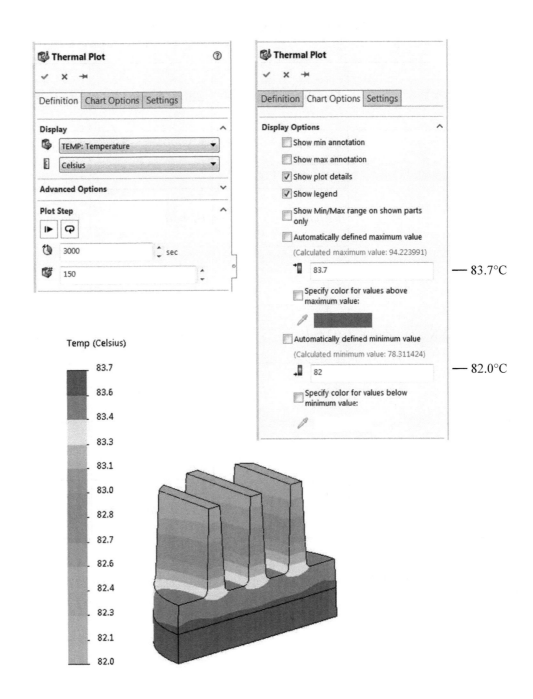

Figure 10-13: Temperature in radiator in the last performed step at time t=3000s.

Use the above selections in Definition and Chart Options tabs.

The temperature plot shows the last performed step number 150 at t=3000s.

The temperature range defined in the plot is 82.0 – 83.7°C.

Summary of studies completed

Model	Configuration	Study Name	Study Type
HEAT SINK TH.sldasm	01		
	01 half	01 half	Thermal
	02		
	02 half	02 half	Thermal
	03		
	03 half	03 half	Thermal
		03 half transient	Thermal

Figure 10-14: Names and types of studies completed in this chapter.

11: Radiative power of a black body

Topics covered

- ❏ Heat transfer by radiation
- ❏ Emissivity
- ❏ Black body
- ❏ Radiating heat out to space
- ❏ Transient thermal analysis
- ❏ Heat power
- ❏ Heat energy

Project description

A graphite ball has a surface area of $1m^2$. It is surrounded by a vacuum and its initial temperature is 0K. At time t=0s, a source of power activates such that 1000W is generated uniformly in the ball volume. We want to know the steady state temperature of the ball's surface and time required to reach that steady state assuming the ball's surface has the emissive properties of a black body.

The objective of this somewhat abstract problem is to study a single body radiating heat out to space without any interference from other objects. This problem has a very simple analytical solution so we will be able to compare numerical and analytical results.

Procedure

Open GRAPHITE BALL and create a thermal study called *01 transient*. Use the study properties to define a study as transient (Figure 11-1).

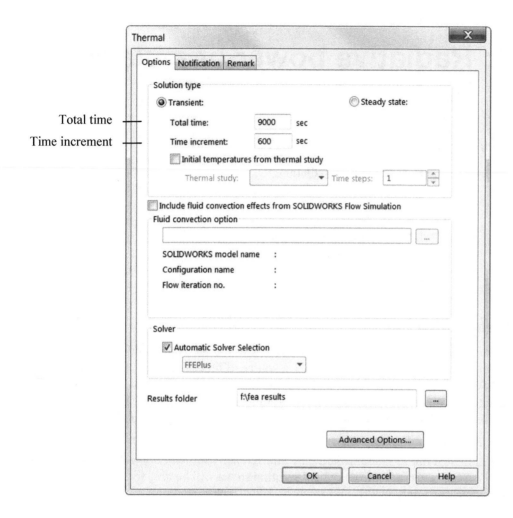

Figure 11-1: Properties of transient thermal study.

Most often, some trial runs are required to define a Total time and Time increment.

As always, in a transient thermal analysis, an initial temperature is required. Follow Figure 11-2 to define it as the absolute zero.

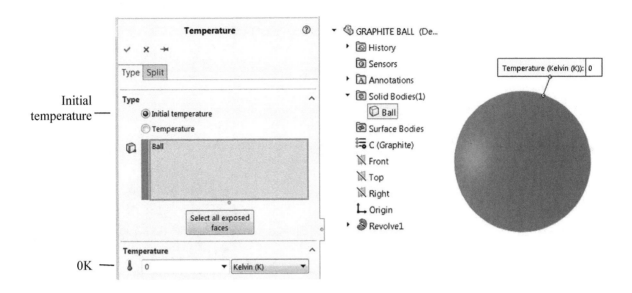

Initial temperature

0K

Figure 11-2: Definition of initial temperature.

From the fly-out menu, select the Solid Body to apply the initial temperature to the entire volume (solid body). Since there is only one Solid Body in this model, this applies the initial temperature to the entire model.

Figure 11-2 shows the definition of initial temperature to the entire volume of the model. Notice that if the surface is selected instead of the volume, then that surface is given the temperature, the rest of the model is at a default of 0K. Therefore, in this specific example, it would not make a difference to apply the initial temperature to the face or to the volume.

Define **Radiation** boundary conditions to the surface of the sphere as shown in Figure 11-3.

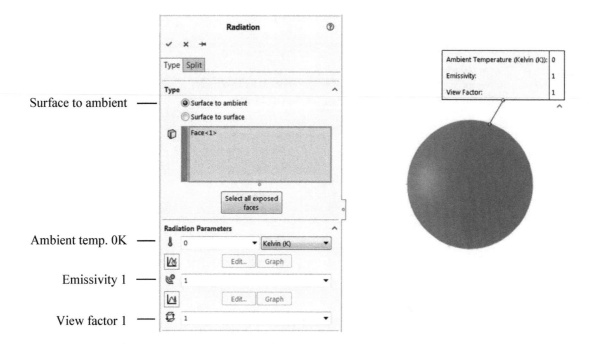

Figure 11-3: Radiation boundary conditions.

Radiation boundary conditions are applied to the face, not to the volume.

An **Emissivity** of 1 means that the ball's face has the emissive properties of a black body. Nothing is better at emitting radiative heat than a black body. A **View Factor** of 1 in combination with a **Surface to ambient** type of radiation boundary condition means that all heat is radiated out to space without interfering with anything.

Define **Heat Power** as shown in Figure 11-4.

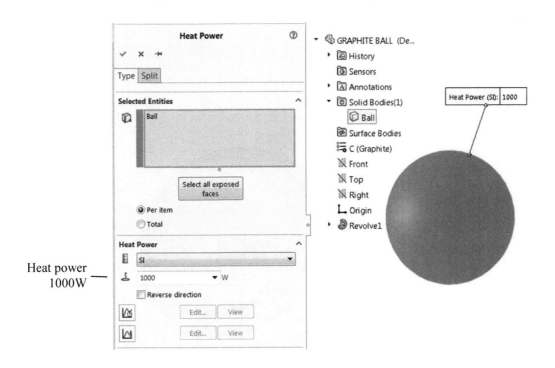

Figure 11-4: Definition of heat power.

From the fly-out menu, select the Solid Body to apply Heat Power to the entire volume of this model.

Mesh the model with a default mesh and run the solution; notice that the solution is completed in 15 steps. Display the **Temperature** results from the last step using a section plot as shown in Figure 11-5.

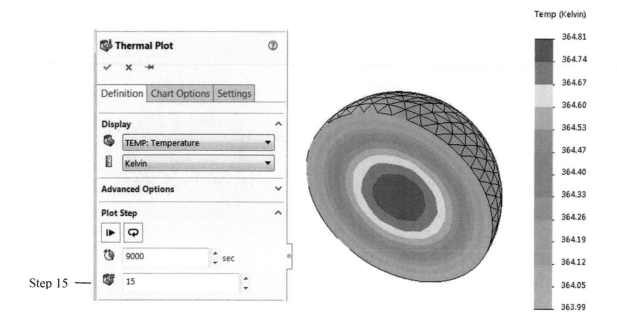

Figure 11-5: Temperature results from step 15.

Mesh is shown on the uncut portion of the model.

The section plot indicates a very low temperature gradient (0.8K) inside the ball where the temperatures are within the range of 364.0K to 364.8K.

To graph a temperature time history, probe any location on the face of the ball.

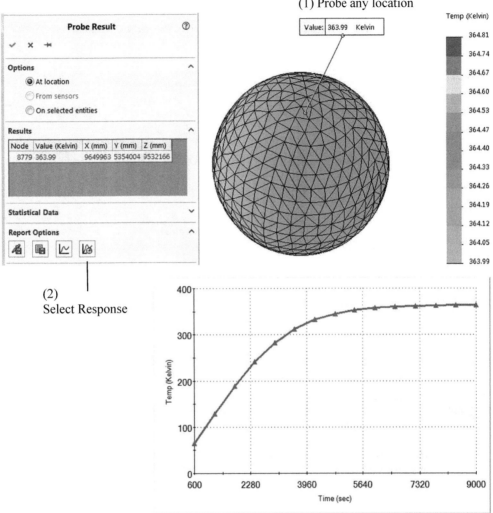

Figure 11-6: The temperature at step 15 (top) and the temperature time history from steps 1-15.

The ball's face temperature is uniform so you may probe any location. Mesh is shown to demonstrate that the probed location snaps to the closest node; in this case, this is a mid-side node.

As the graph in Figure 11-6 indicates, the steady state temperature has been practically reached after about 7000s.

Review the **Heat Power** at step 15 and **Heat Energy** produced during the 9000s (Figure 11-7).

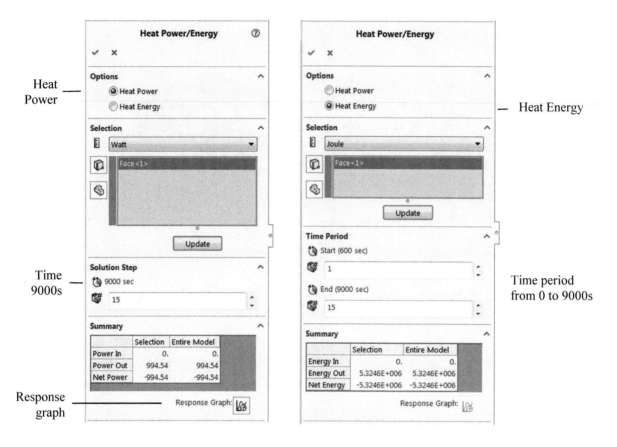

Heat Power emitted by the surface at step 15 (9000s)

Heat Energy radiated out during 9000s

Figure 11-7: Heat Power at step 15 and total Heat Energy during 9000s. Select the face in both windows.

After 9000s the face is emitting 995W, the total energy emitted by radiation during 9000s is 5.325MJ. Review Response Graph in Heat Power window.

Use windows as shown in Figure 11-8 to collect information about heat power radiated out at the end of each step. The summary is shown in Figure 11-9.

Step 1: 690W   Step 10: 926W   Step 15: 995W

<u>Figure 11-8: Energy radiated out by the ball surface during selected steps.</u>

Remember to click Update every time the Solution Step is changed. Review response Graph in any of the above windows.

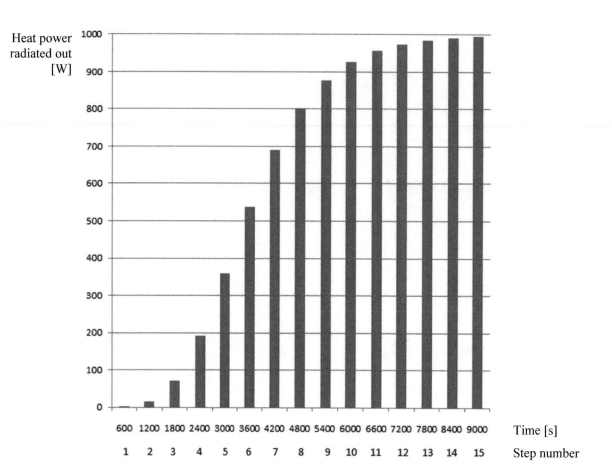

Figure 11-9: Heat power radiated out by the ball surface as function of time.

Compare this graph with the temperature time history graph in Figure 11-6.

The same graph may be created automatically by selecting **Response Graph** in any of the windows shown in Figure 11-8.

Very simple model geometry makes it easy to compare the results of the numerical simulation with analytical results. Review model and results to notice that:

The surface area of the ball is $1m^2$

The emissivity of the ball face is $\varepsilon = 1$ (it is a black body)

Stefan-Boltzmann constant $\sigma = 5.67 \times 10^{-8}$ W/m²/K⁴ (see chapter 1)

Heat flux $q = 995W/1m^2 = 995W/m^2$ (Figure 11-8)

Using the above, the surface temperature T_S can be found from the Stefan-Boltzman law:

$$q = \varepsilon \sigma T_S^4$$

$$T_S = \left(\frac{q}{\varepsilon\sigma}\right)^{0.25} = \frac{995}{1 \times 5.67 \times 10^{-8}} = 364K$$

This result is the same as the numerical result $T_S = 364K$ shown in Figure 11-5. If results were truly steady state, we could use a heat flux of $1000W/m^2$ from the input data.

We will now review energy balance.

Specific heat of graphite: $c = 44$ J/kg/K (read from material properties)

Mass of graphite ball: 210.64kg (measure this in the SOLIDWORKS model)

Energy used to raise the temperature of the graphite ball from 0K to 364K:

$$E_C = mc\Delta T = 210.64 \times 44 \times 364 = 3374kJ$$

Energy radiated out from the ball's surface (Figure 11-7):

$$E_R = 5325kJ$$

Total energy:

$$E = E_C + E_R = 3374kJ + 5325kJ = 8699kJ$$

This result is within 4% of the total energy calculated as:

$$E = P \times t = 1000W \times 9000s = 9000kJ,$$

where P is heat power and t is the duration of the analyzed process.

Summary of studies completed

Model	Configuration	Study Name	Study Type
GRAPHITE BALL.sldprt	*Default*	*01 transient*	Thermal

Figure 11-10: Names and types of studies completed in this chapter.

12: Radiation of a hemisphere

Topics covered

- ❏ Heat transfer by radiation
- ❏ Emissivity
- ❏ Radiating heat out to space
- ❏ View factors
- ❏ Heat power

Project description

A nickel hemisphere is heated in its volume with a heat power of 800W. All faces are insulated except for the concave face that exchanges heat by radiation with a large enclosure at 30°C. The concave face emissivity is 0.4. The hemisphere is in vacuum. We need to find the steady state temperature of the hemisphere.

Procedure

Open model HEMISPHERE and create a thermal study called *01 steady state* (Figure 12-1).

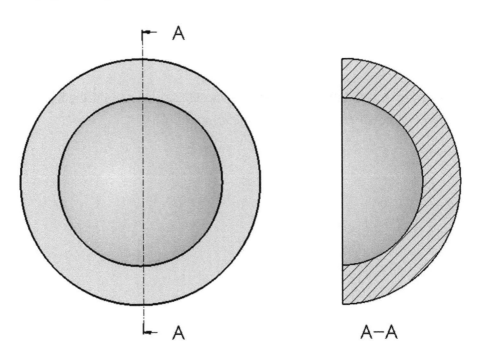

Figure 12-1: HEMISPHERE model in front and section views.

The concave face radiates heat out to open space characterized by an ambient temperature of 30°C. The other two faces are insulated. Heat transfer takes place only through the concave face. The above views use the third angle projection.

Define **Heat Power** as shown in Figure 12-2.

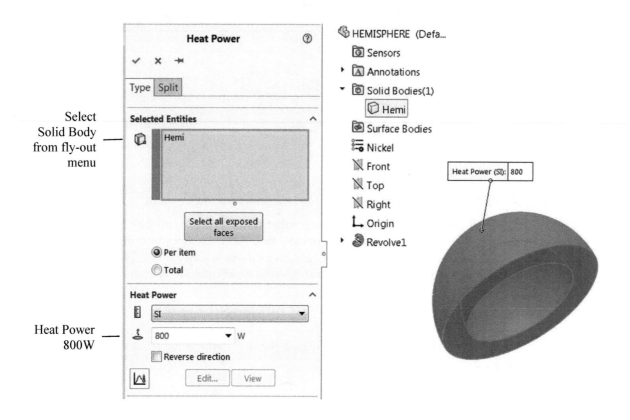

Figure 12-2: Heat Power is generated in the entire volume of the model.

Select Solid Body Hemi from the fly-out menu.

Define **Radiation** boundary conditions as shown in Figure 12-3.

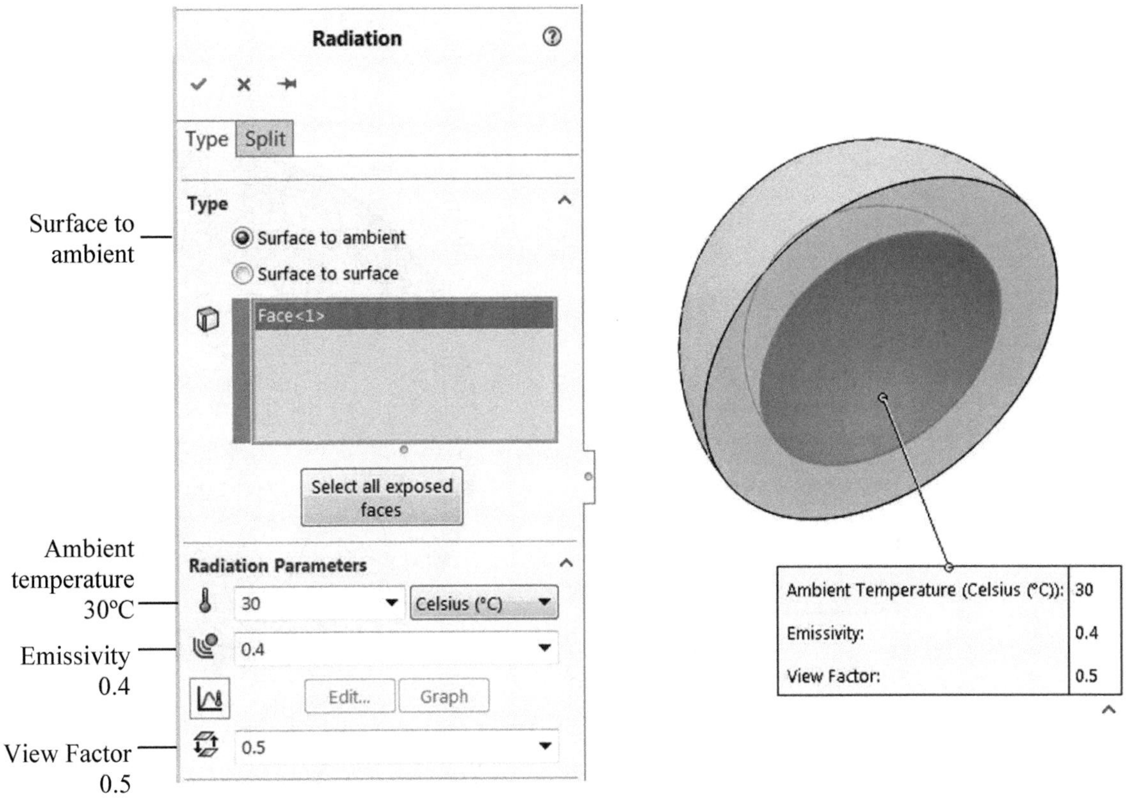

Figure 12-3: Radiation boundary conditions.

Radiation boundary conditions are defined on the concave face.

You may conceptualize **Ambient temperature** as the temperature of a large enclosure in which the HEMISPHERE is located: Figure 1-7 (2). **Emissivity** is a measure of how well the face emits radiation in comparison to a black body, which has the emissivity equal to 1. **View Factor** is a measure of what portion of radiation is radiated out directly to space. If radiation boundary conditions were **Surface to ambient** with a **View Factor** equal to 1, then all power would be radiated out directly to space as was the case with the GRAPHITE BALL in chapter 11. The **View Factor** of 0.5 used in the HEMISPHERE accounts for the fact that this concave face can "see itself"; some of the heat radiated out hits this face again before being finally radiated out to space.

Mesh the model using a **Curvature based mesh**; use element size 15mm as shown in Figure 12-4.

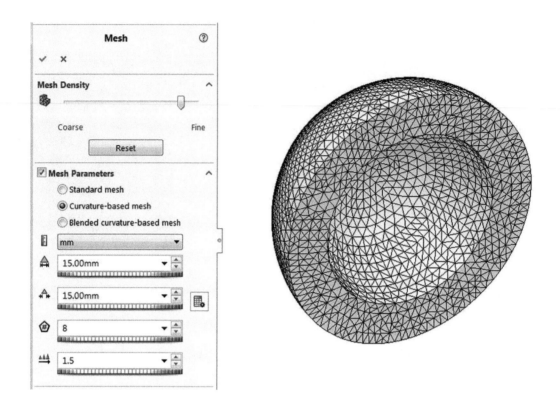

Figure 12-4: Mesh parameters are all on defaults.

A curvature based mesh is a natural choice for this type of geometry.

Obtain a solution and show a **Temperature** plot (Figure 12-5).

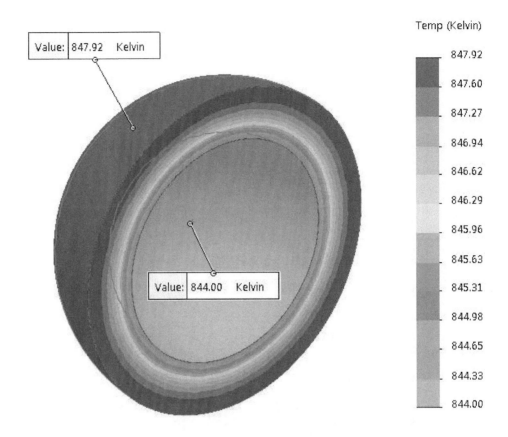

Figure 12-5: Temperature result plot; probed is the temperature of the outside convex and of the inside (concave) face.

In radiation problems, temperature is often expressed in Kelvins.

Notice a low temperature gradient in the model due to a high thermal conductivity of nickel.

Follow steps in Figure 12-6 to review **Heat Power** results and verify that all power generated in the volume to HEMISPHERE is radiated out by the concave face where the radiation boundary conditions have been defined.

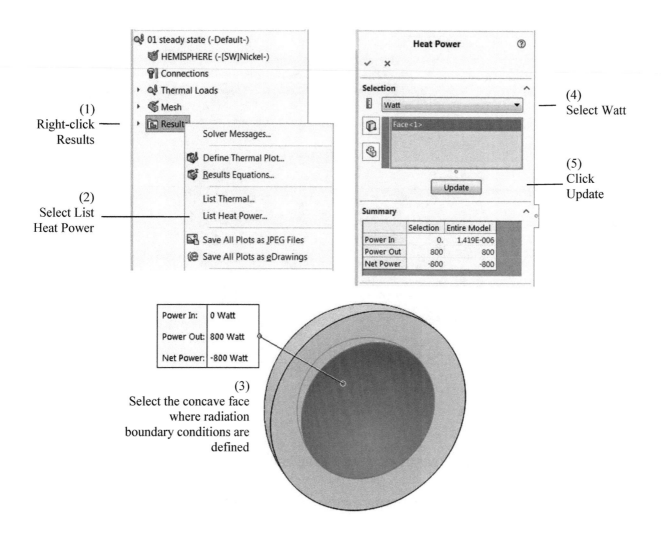

Figure 12-6: Heat Power radiated out is 800W.

The generated heat leaves the HEMISPHERE through the face where the radiation boundary conditions are defined. It is equal to the heat power generated in the volume of the model.

Summary of studies completed

Model	Configuration	Study Name	Study Type
HEMISPHERE.sldprt	*Default*	*01 steady state*	Thermal

Figure 12-7: Names and types of studies completed in this chapter.

Notes:

13: Radiation between two bodies

Topics covered

- ❏ Heat transfer by radiation
- ❏ Emissivity
- ❏ Radiating heat out to space
- ❏ View factors
- ❏ Heat power
- ❏ Closed system
- ❏ Open system

Project description

Heat power is generated in the volume of one solid body and is radiated out to the second body, which dissipates it by convection. This example may have little relevance to any practical problem but it allows studying different options in radiation heat transfer available in **SOLIDWORKS Simulation**.

Procedure

Open assembly model BLOCKS (Figure 13-1) and create a thermal study *01 surface to surface closed system.*

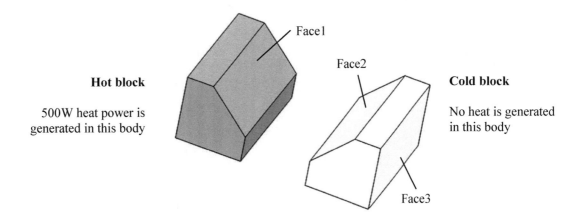

Hot block

500W heat power is generated in this body

Cold block

No heat is generated in this body

Figure 13-1: BLOCKS assembly consists of two identical solid bodies.

There are three mechanisms responsible for heat transfer in this model: conduction, radiation and convection. Heat moves by conduction inside the solid bodies. Face 1 and Face 2 exchange heat by radiation. Face 3 dissipates heat by convection.

In **Simulation** heat transfer between two or more detached bodies can only be exchanged by radiation. This is because convection models heat exchange between a face and surrounding fluid, not between two faces.

Apply **Heat Power** as shown in Figure 13-2.

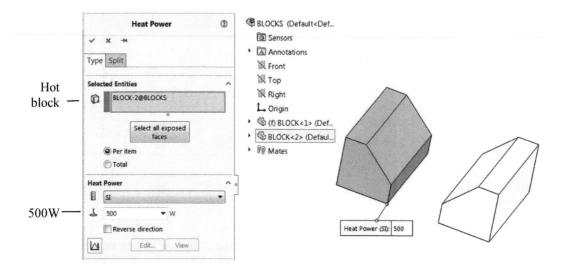

Figure 13-2: 500W Heat Power is applied to the left block.

Select the block from the fly-out menu.

Apply **Radiation** boundary conditions as shown in Figure 13-3.

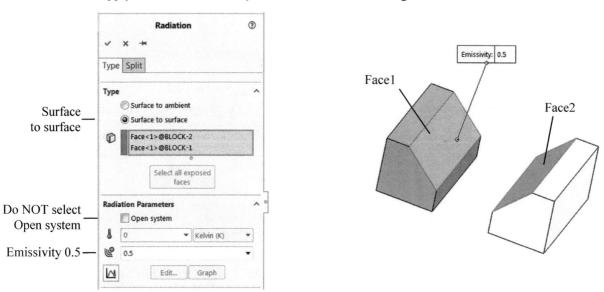

Figure 13-3: Radiation boundary conditions. Heat is exchanged by radiation between the selected Face1 and Face2.

Emissivity 0.5 applies to both faces.

It is important to comment on the **Radiation** boundary conditions defined in Figure 13-3. Without selecting **Open system** we assume that all heat radiated out by face 1 travels without any loss to face 2. This is of course unrealistic but the objective of this exercise is to demonstrate differences between **Radiation** boundary conditions in **Open** and **Closed** systems.

So far we have established a mechanism of heat flow by radiation between two bodies. Now we have to define how heat exits the system. Define **Convection** boundary conditions as shown in Figure 13-4.

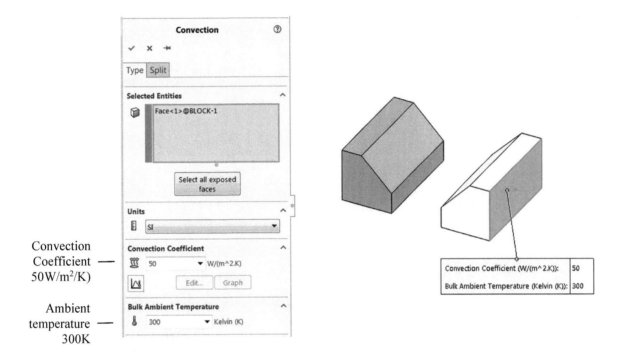

Convection Coefficient — 50W/m²/K)

Ambient temperature — 300K

Figure 13-4: Convection boundary conditions, the selected face exchanges heat with fluid.

Convection symbols are not shown in this illustration.

The face where the **Convection** boundary condition is defined must be in contact with some fluid to be able to exchange heat by convection. The faces with **Radiation** boundary conditions are exposed to a vacuum but no heat is radiated out to space because **Open system** is deselected

No thermal boundary conditions are defined on the remaining faces meaning that no heat exchange takes place there.

Mesh the model with a default mesh and run the solution. Review the **Temperature** plot as shown in Figure 13-5.

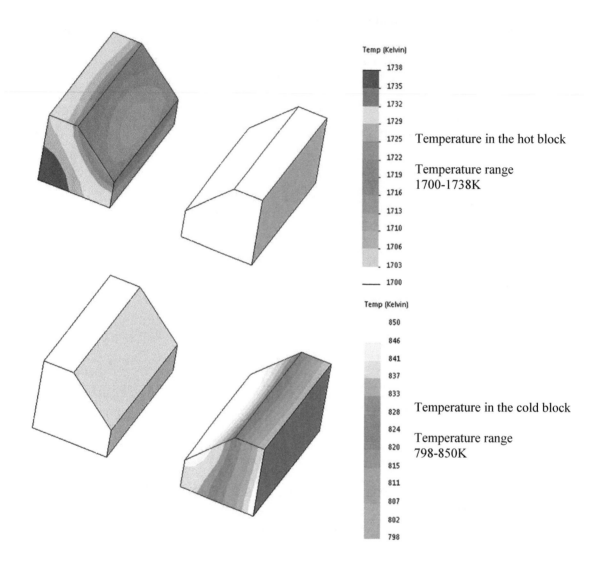

Figure 13-5: Temperature results.

Two plots with different temperature ranges are used to show temperature distribution in the hot and cold blocks.

Temperature results are presented in Figure 13-5 using two plots with different temperature ranges: 1700-1738K to show temperature distribution in the block radiating out heat and 798-850K to show temperature distributions in the block receiving heat by radiation.

Review **Heat Flux** results using a vector plot as shown in Figure 13-6.

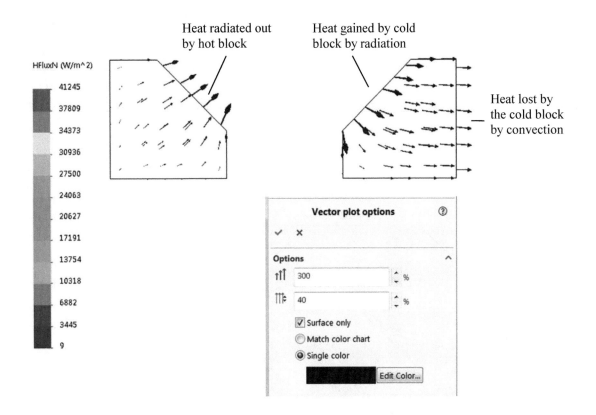

Figure 13-6: Heat Flux results.

The vector plot offers a convenient way to analyze the path of heat flow. Vector plot options window is shown.

To verify that the entire heat generated in the hot block indeed travels through a vacuum to the cold block, we may probe **Heat Power** on any of the three faces that participate in heat exchange as shown in Figure 13-7.

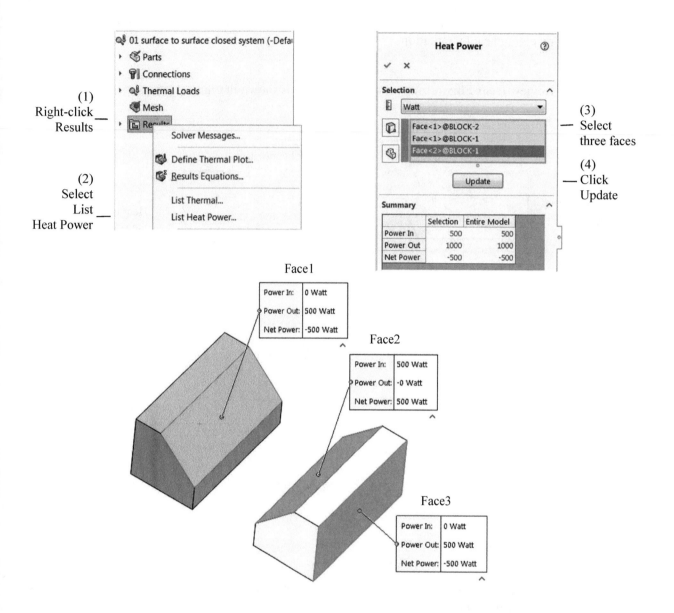

Figure 13-7: Heat Power results.

Heat Power is probed on all faces where thermal boundary conditions have been defined.

The minus sign by the **Net Power** indicates the net loss of 500W among the three faces.

This completes the analysis carried under the assumption of a closed system. Now copy the study *01 surface to surface closed system* into study *02 surface to surface open system*. The only difference between these two studies is that the radiation boundary condition now specifies **Open system** (Figure 13-8).

Closed system —

Open system —

Ambient temperature

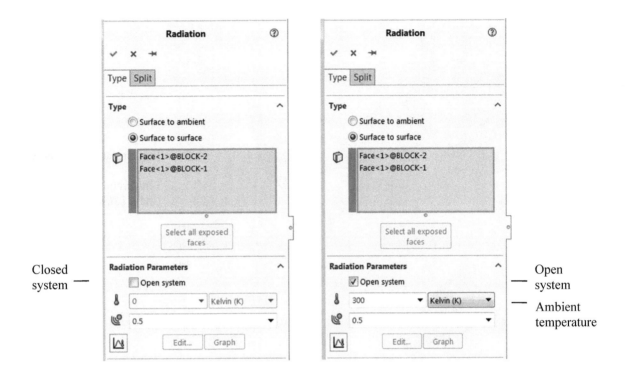

Study *01 surface to surface closed system*   Study *02 surface to surface open system*

Figure 13-8: Radiation boundary conditions in two studies.

Open system allows the heat to be radiated out to space.

Notice that while specifying **Open system** we also have to define **Ambient Temperature** (here 0K) because heat is now radiated out to space. **Ambient temperature** is the temperature of a distant enclosure inside in which the two blocks are located as shown in Figure 1-7 (2).

Temperature and **Heat Flux** results in **Open system** are shown in Figures 13-9 and 13-10.

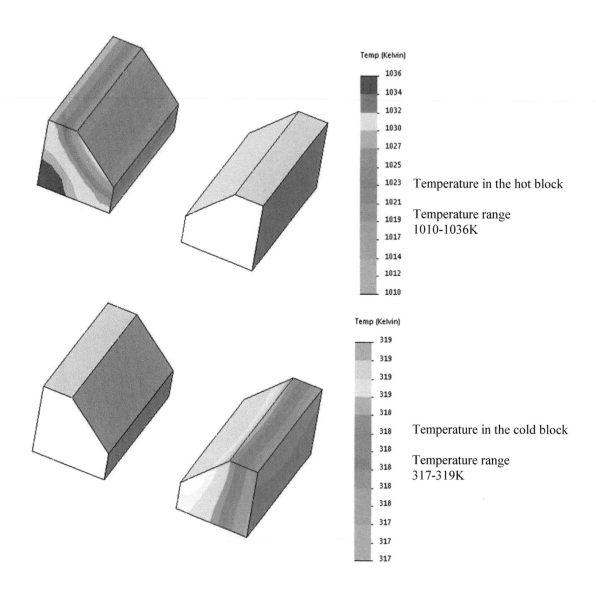

Temperature in the hot block

Temperature range
1010-1036K

Temperature in the cold block

Temperature range
317-319K

Figure 13-9: Temperature results in Open system.

Two plots with different temperature ranges are used to show temperature distribution in the hot and cold blocks.

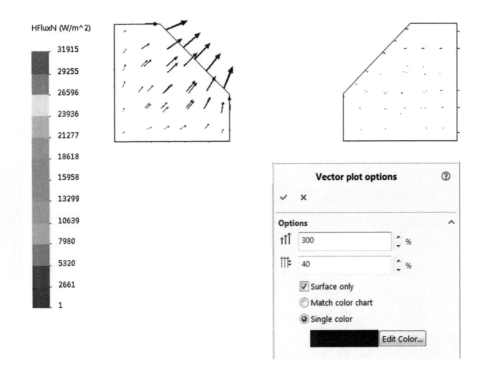

Figure 13-10: Heat Flux results in Open system.

Heat Flux vectors are barely visible in Cold block because very little heat reaches to the Cold block. Most of it is lost to the surrounding space.

Follow the same steps as shown in Figure 13-7 to probe **Heat Power** and find out how much heat is transferred from the hot block to the cold block and how much is radiated out to space (Figure 13-11).

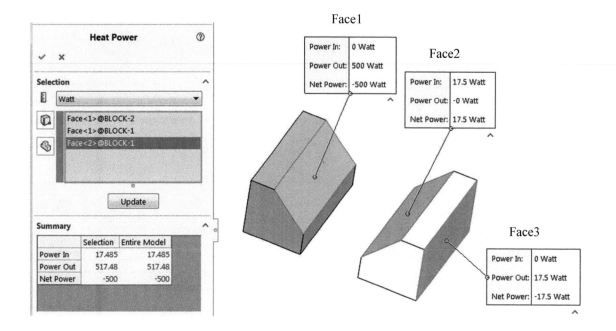

<u>Figure 13-11: Heat Power results in an Open system.</u>

Out of 500W leaving Face1, only 17.5W reaches Face2.

As Figure 13-11 shows, 500W - 17.5W = 482.5W is radiated out to space. The 17.5W of heat that reaches the Cold block travels by conduction through the Cold block to the face where the **Convection** boundary condition is defined. Upon reaching it, it is dissipated by convection to the surrounding fluid.

To illustrate an important difference between the study with **Closed system** and **Open system**, delete or suppress the **Convection** boundary condition in study *01 surface to surface closed system* and try running it. This results in an error (Figure 13-12) because in a closed system, in the absence of **Convection**, a mechanism of heat transfer does not exist.

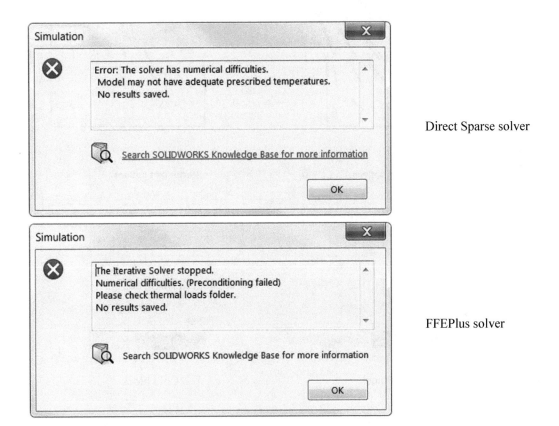

Direct Sparse solver

FFEPlus solver

Figure 13-12: An error message is displayed when the solution is attempted with a Closed system and no Convection boundary conditions.

Direct Sparse solver and FFEPlus solver display different error messages.

If you suppress **Convection** using an **Open system**, the solution will be completed. This is because Face1 and Face2 can now exchange heat not only between themselves, but also can radiate it out to space (Figure 13-13).

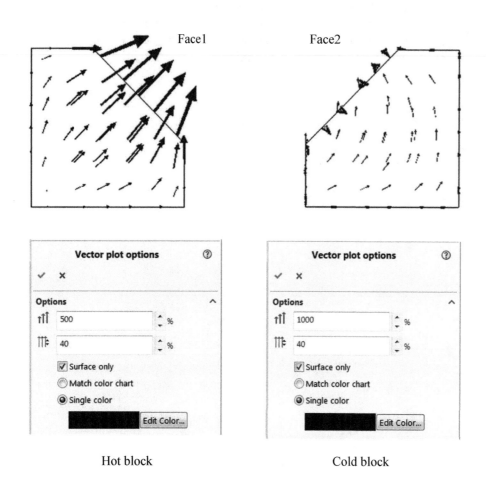

Hot block   Cold block

Figure 13-13: In the absence of convection in the Open system, heat enters and exits the Cold block through Face2.

To improve clarity of vector plots, different Vector options are used for the hot block and for the cold block.

Summary of studies completed

Model	Configuration	Study Name	Study Type
BLOCKS.sldasm	*Default*	*01 surface to surface closed system*	Thermal
		02 surface to surface open system	Thermal

Figure 13-14: Names and types of studies completed in this chapter.

Notes:

14: Heat transfer with internal fluid flow

Topics covered

❏ Introduction to Flow Simulation

❏ Using Flow Simulation for finding convection coefficients in internal fluid flow

❏ Interfacing between Flow Simulation and Thermal analysis

❏ Interfacing between Flow Simulation and structural (Static) analysis

Project description

In the preceding chapters, we introduced thermal analysis as a tool of heat transfer analysis in solids. On many occasions we pointed out that heat exchange between a solid body and a surrounding fluid is governed by convective boundary conditions, but the fluid flow itself is not modeled in **SOLIDWORKS Simulation**. To define heat exchange between the face of a solid and the surrounding fluid, we would define a convection coefficient and a reference (ambient) temperature.

In this chapter, we use a very different approach; we will find the convective boundary conditions (convection coefficients and ambient temperature) by simulating fluid flow. We will then use those convective boundary conditions to analyze heat transfer in a solid body the way we were doing that in chapters 1-13.

Numerical simulation of fluid flow belongs to a discipline called **Computational Fluid Dynamics (CFD)**. In **SOLIDWORKS** it has been implemented as an add-in called **Flow Simulation** which is different from **Simulation**. **Flow Simulation** is a topic for a separate book; we will introduce it here only in a limited scope as required to use the results of **Flow Simulation** in a **Thermal** and **Static** or **Nonlinear** analyses, which belong to **Simulation**.

We will be using two add-ins to **SOLIDWORKS**: **Simulation** and **Flow Simulation**. These similar and somewhat confusing names require some introduction. First, let us agree on the following name convention: **Simulation** will mean the add-in in called **SOLIDWORKS Simulation**; this is what we have been using exclusively until now. **Flow Simulation** will mean another add-in used for **CFD**.

Simulation is based on a numerical technique called **Finite Element Method** which we customarily call **FEA**. **Simulation** is used for structural and thermal analyses of solid bodies. **Flow Simulation** is based on a numerical technique called **Finite Volumes (FV)**. It is used for the analysis of fluid flow, which may, but does not have to include heat transfer.

If heat transfer is not considered in the analyzed problem, then **Flow Simulation** is concerned only with the space where fluid flow occurs. This may be a cavity inside a solid and/or space around a solid.

If heat transfer is included in the analysis of fluid flow, then **Flow Simulation** models the heat transfer both in fluids and in solids. Convection coefficients governing heat transfer between fluids and solids are found from fluid flow conditions. This is very different from a **Thermal** analysis with **Simulation** where the convection coefficient must be known.

As we can see, there is an overlap between **Simulation** and **Flow Simulation**: both can do thermal analysis of solids. However, **Flow Simulation** is not intended for thermal analysis of solids; it does it only to the extent necessary to solve thermal conditions present in the analyzed fluid flow.

Flow Simulation can be interfaced in different ways with **Thermal** and structural (**Static**, **Nonlinear**) analyses available in **Simulation** (Figure 14-1).

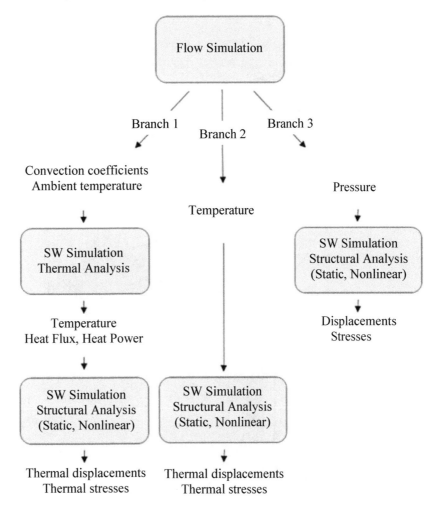

Figure 14-1: Interfacing between Flow Simulation and SOLIDWORKS Simulation.

If only thermal displacements or stresses are required, it is preferred to follow the right branch (Branch 2). Branch 3 is outside of the scope of this book.

We should point out that **Flow Simulation** is a tool of flow analysis and is not intended for thermal analysis in solids. It is preferable to use **Flow Simulation** to find convection coefficients, transfer them to a **Thermal** analysis, then use the **Thermal** analysis to find temperature and everything else describing heat transfer. This may be followed by the analysis of thermal stresses in a **Static** or **Nonlinear** analysis. This process is shown in the left branch in Figure 14-1.

A simple assembly model ELBOW (Figure 14-2) will serve to introduce the **Finite Volumes** method used in **Flow Simulation**.

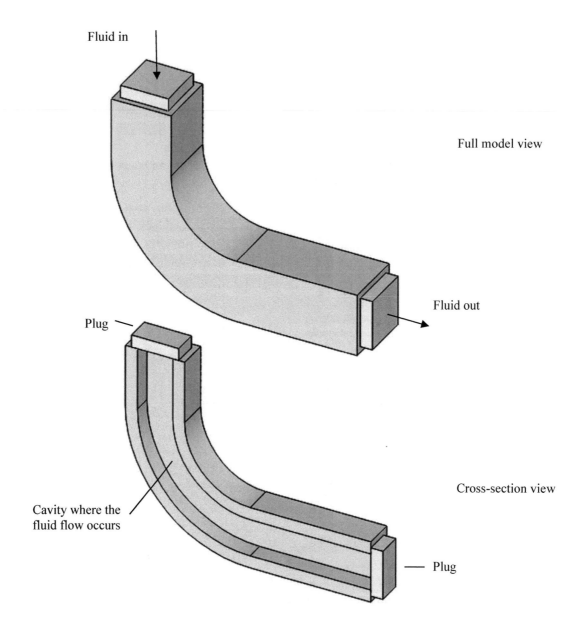

Fluid in

Full model view

Fluid out

Plug

Cross-section view

Cavity where the
fluid flow occurs

Plug

Figure 14-2: ELBOW assembly model consists of a square tube and two plugs.

Two plugs make the model "water-tight." Cross section view is used to show the cavity where the fluid flow occurs.

The flow of fluid happens here in the space (cavity) inside the solid model, not in the solid model itself. Definition of the space when fluid flow occurs requires plugging up the model as shown in Figure 14-2. In fluid flow analysis, solid geometry is used only to define the space where fluid flow occurs (inside the cavity, around a solid or both). In our case the flow is inside; therefore, only the inside faces of the cavity matter for analysis. As we mentioned earlier, solid geometry is considered in flow analysis only if heat transfer in fluids and solids is specified in the analysis settings. The combined heat transfer in fluids and solids is called **conjugate heat transfer**. It is classified in **Flow Simulation** as an advanced topic. To introduce **Flow Simulation** gradually, the ELBOW example won't include heat transfer.

Using the **Finite Volumes,** the space where fluid flow occurs is discretized into cubes parallel to the global coordinate system. If heat transfer in solids is requested in analysis, then solids are also discretized. A **Finite Volumes** mesh, most often referred to as a grid, can only be shown in cross-sections as illustrated in Figure 14-3.

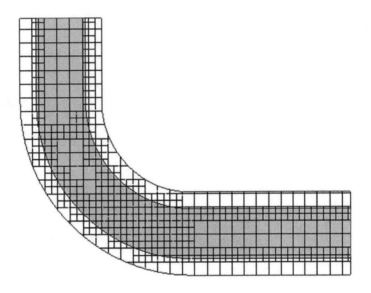

Figure 14-3: Cross section through the Finite Volume grid.

Cells modeling fluid are shaded in this illustration. Cells modeling solid body (white) are here only because heat transfer in solids has been requested prior to grid generation. Partial cells modeling solids and fluid are located along curvilinear boundaries, which require a more refined grid of orthogonal cubes to approximate the curvature; therefore, smaller cells are located along the rounds.

Solid cells will not be present in the analysis that follows. They are shown here for illustration purposes only.

The grid cells may be classified into three types: fluid, solid and partial. Partial cells are located on boundaries between fluid and solid and include both solid and fluid volume. The size of the cubes depends on the geometry and on the requested level of solution accuracy. 3D grid refinement is conducted by splitting a cell into eight smaller cells.

Parameters characterizing fluid flow and heat transfer are evaluated in the middle of a cell so modeling of a linear temperature gradient in a solid wall requires at least two layers of cells.

As we said before, the ELBOW model is an introductory exercise in fluid flow analysis with no thermal conditions. First, make sure that **Flow Simulation** is an active add-in (Figure 14-4) and is showing as a tab in the **Command Manager**.

Figure 14-4: Activate SOLIDWORKS Flow Simulation and review Flow Simulation tab.

Simulation add-in and Flow Simulation add-in should be selected.

Start a new **Flow Simulation** project using a **Wizard** as shown in Figure 14-5. The **Wizard** will walk you through steps necessary to set up a flow analysis project.

(2) Select Wizard

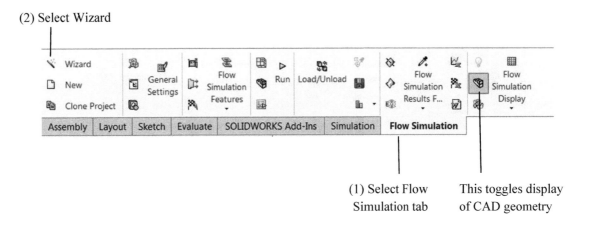

(1) Select Flow Simulation tab

This toggles display of CAD geometry

Figure 14-5: Starting a Flow Simulation Project using a Wizard.

The Wizard will walk you through the steps necessary to define a project, which is equivalent to a "Study" in Simulation.

After exiting Wizard, experiment with toggling CAD geometry display.

Steps in the project definition are shown in Figures 14-6 through 14-12. Steps are described in captions of illustrations and indicated by arrows in the **Navigator** window.

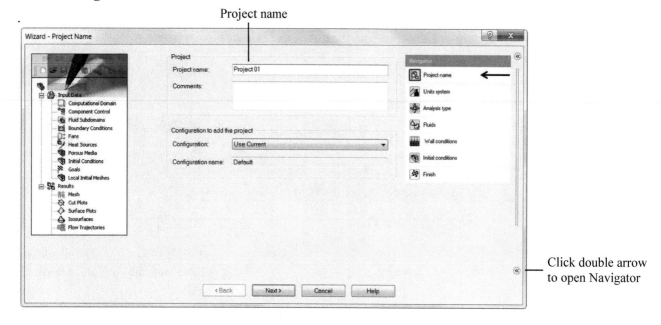

Figure 14-6: Step 1: Project name; enter project name Project 01.

The project windows are more informative if you keep the Navigator window open. The window on the right lists all the steps performed during project definition. Click Next or Back to move through the steps.

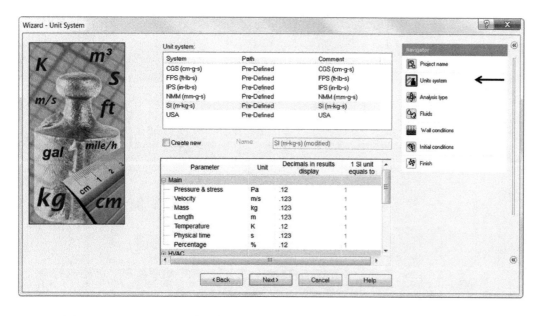

Figure 14-7: Step 2: Selection of units.

Accept the default SI system of units.

Select Internal

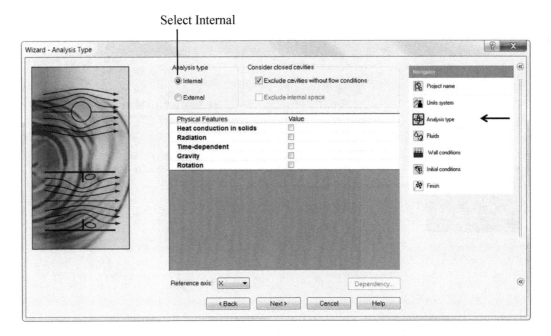

Figure 14-8: Step 3: Analysis type.

Select the Internal type of analysis; do not select Heat conduction in solids in this exercise.

Open Pre-Defined sub-folder
in liquids folder, Select Water

Click Add to add Water
as the default fluid

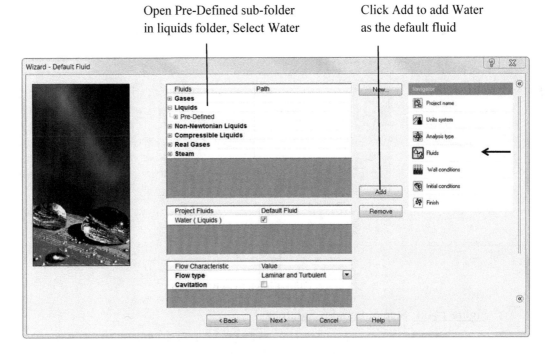

Figure 14-9: Step 4: Default Fluid.

Select Water from the list of pre-defined fluid.

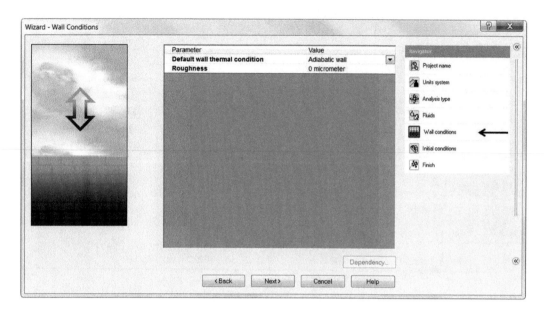

Figure 14-10: Step 5: Wall conditions.

Adiabatic wall means no heat exchange; do not change anything here.

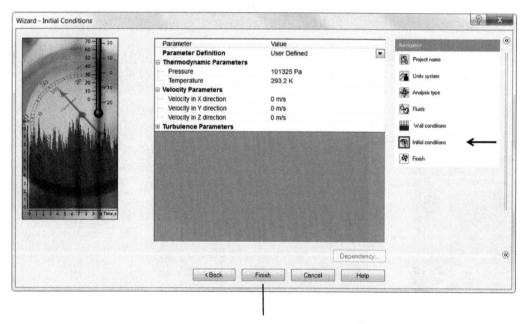

Click Finish to exit Wizard

Figure 14-11: Step 6: Initial Conditions.

Do not change anything here. The example we are setting up is a steady state fluid flow; velocity parameters will be found by solver.

When the **Wizard** is closed, **Flow Simulation** creates a Project window with several folders in it. Right-click on *Mesh* folder to open Global Mesh Settings window (Figure 14.2)

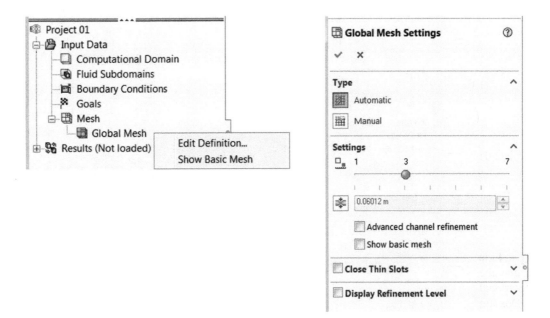

Figure 14-12: Global Mesh Settings window.

Accept the default resolution level 3. This controls the level of grid refinement performed during the solution process. It is comparable to mesh refinement in FEA.

A **Flow Simulation** project tab has been added and the CAD model view now shows the **Computational Domain**. Notice that it does not include solid walls because we have not selected heat transfer in solids.

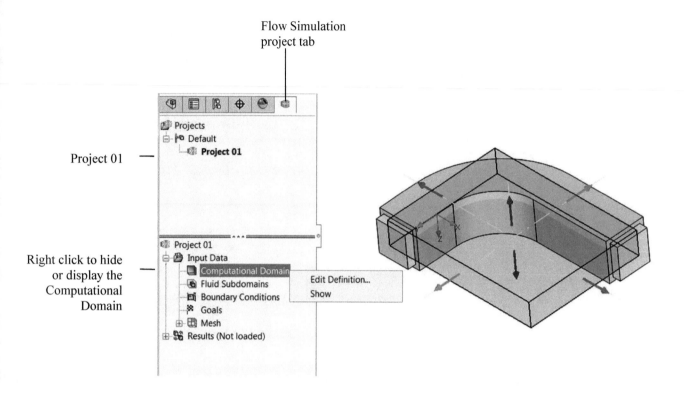

Flow Simulation
project tab

Project 01

Right click to hide
or display the
Computational
Domain

Figure 14-13: The computational domain is aligned with the global coordinate system.

Practice hiding and showing the CAD model and the Computational Domain.

Familiarize yourself with the **Flow Simulation** Command Manager, which is now active.

We will now set up conditions in the inlet and outlet defining fluid flow. Right-click **Boundary Conditions** and select **Insert Boundary Conditions**, then follow the steps in Figure 14-14.

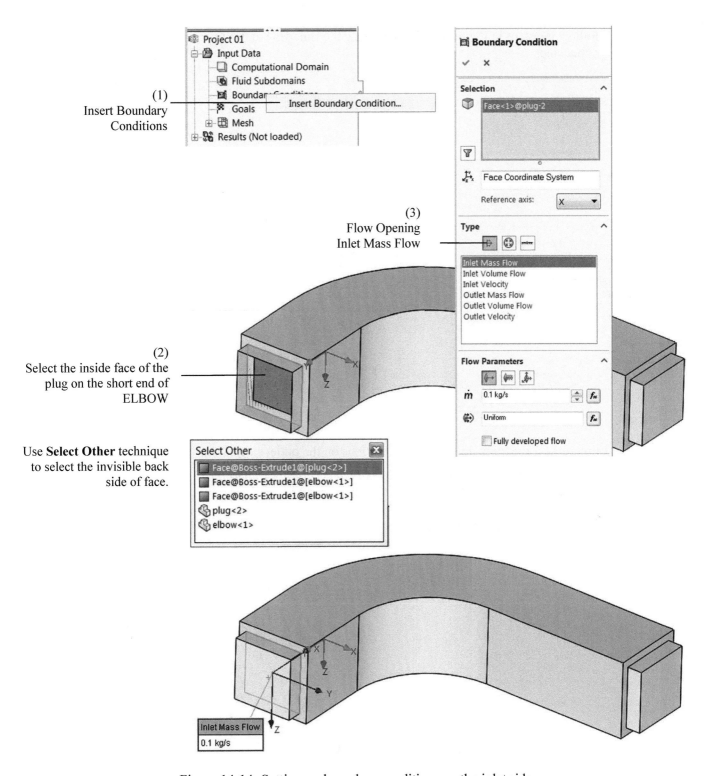

Figure 14-14: Setting up boundary conditions on the inlet side.

Notice that a call-out is added to the face where the boundary condition has been defined. Instead of using Select Other technique, you may use cross-section view to select the desired face directly.

Follow the steps in Figure 14-15 to complete the fluid flow definition on the outlet side.

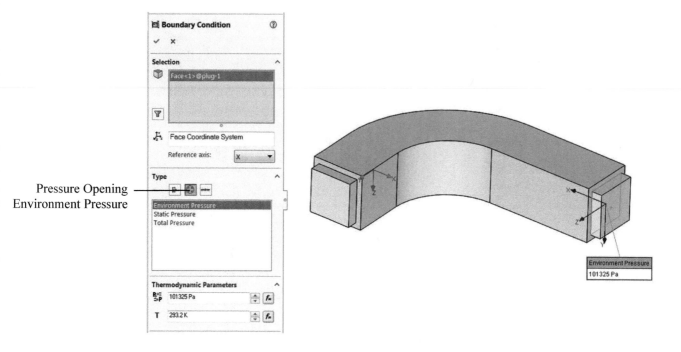

Figure 14-15: Setting up boundary conditions on the outlet side.

Accept the default pressure and the default temperature. Temperature has no importance in this analysis.

The last step before running the solution is the setup of **Goals** for the project. These are flow parameters, which are of interest, and, therefore, we ask the solver to consider them when it performs the solution convergence check.

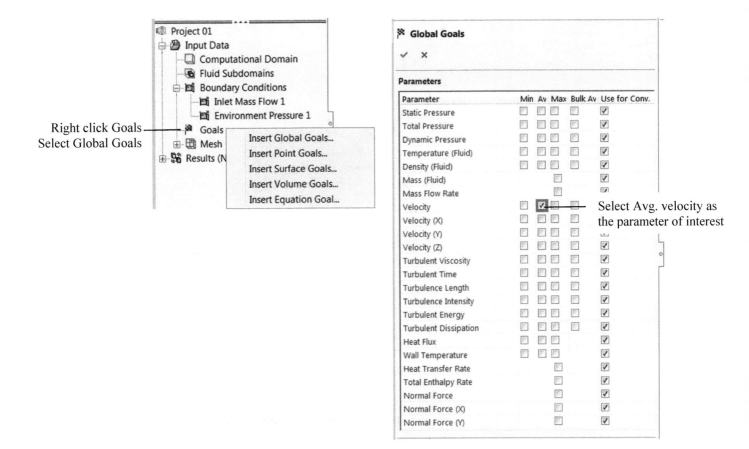

Figure 14-16: Definition of solution Goals.

Only one Global Goal is selected as parameter of interest.

The flow analysis in ELBOW is for demonstration purpose only; therefore, we define only one **Goal**.

Right-click **Project 01** in either place shown in Figure14-17 and select **Run**.

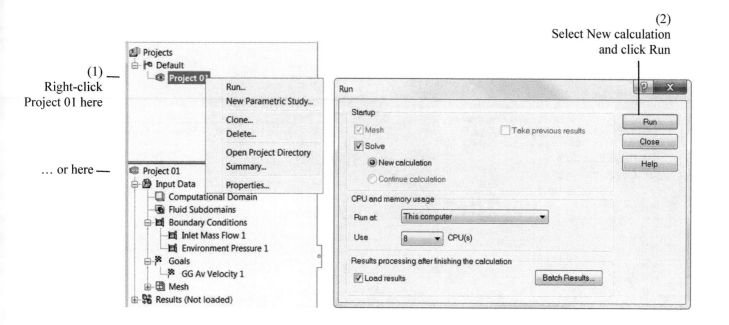

Figure 14-17: Running the solution of Project 01.

Starting the solution requires two steps as shown above.

Solution progress may be monitored using the **Solver** window shown in Figure 14-18.

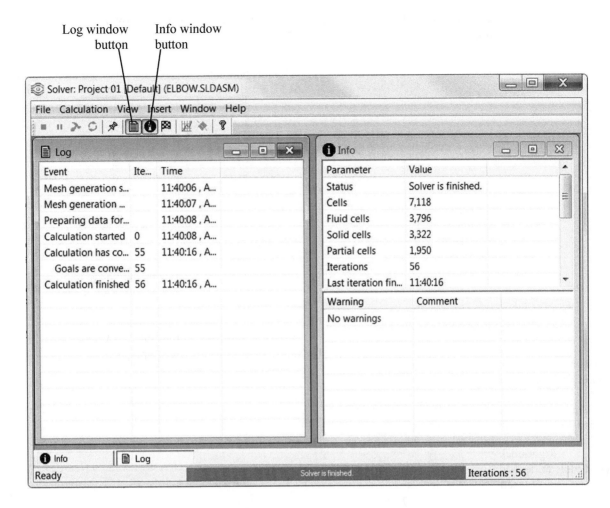

Figure 14-18: Solver window.

Use Info and Log windows to monitor the solution progress. These windows are activated using the buttons shown above.

Closing the **Solver** window allows you to access the results of the flow analysis (Figure 14-19).

Use this button to Load/Unload results

Use this button to turn on/off light controlling brightness of plots

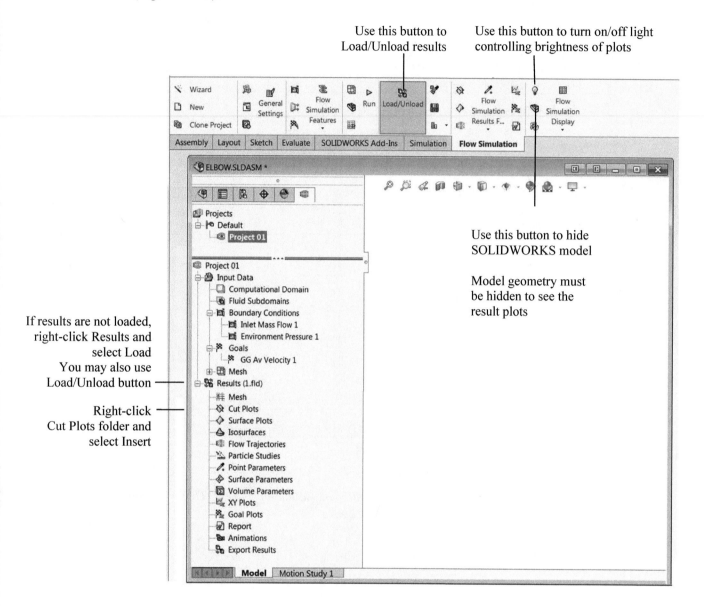

If results are not loaded, right-click Results and select Load You may also use Load/Unload button

Right-click Cut Plots folder and select Insert

Use this button to hide SOLIDWORKS model

Model geometry must be hidden to see the result plots

Figure 14-19: Result folders.

Right-click the desired folder and select Insert to add a plot of the desired type.

In this introductory exercise, we limit the analysis of results to a quick review of pressure (Figure 14-20) and velocity plots (Figure 14-21).

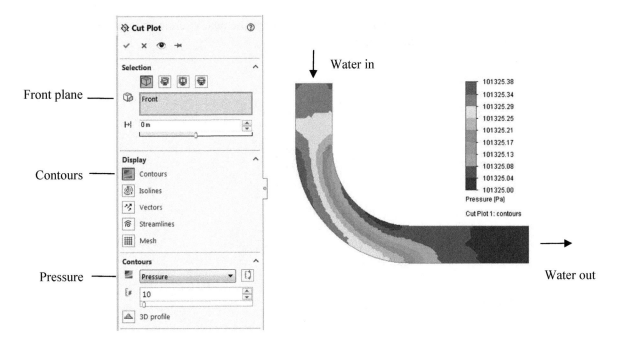

Figure 14-20: Pressure plot; a very low pressure drop can be seen.

To construct this plot, make selections as indicated.

Hide the above plot before constructing the next one.

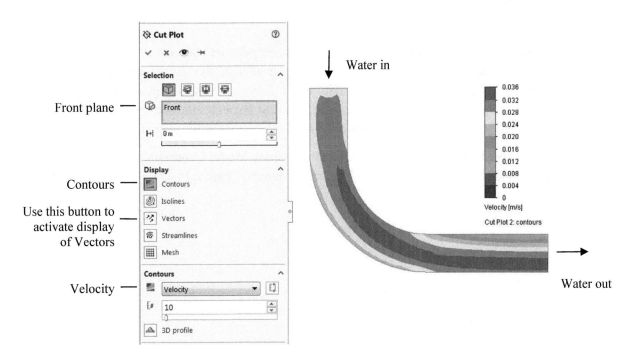

Figure 14-21: A velocity plot illustrates a non-uniform velocity distribution caused by the curvature of ELBOW.

To construct this plot, make selections as indicated.

To complete this review and prepare yourself for the next exercise construct a mesh plot as shown in Figure 14-22.

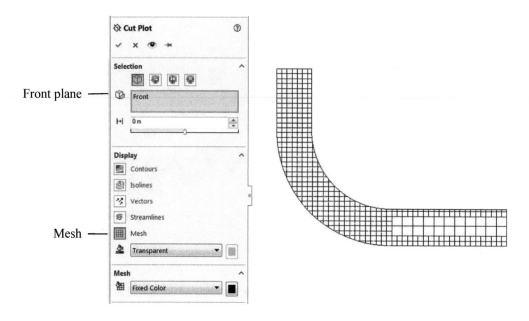

Figure 14-22: Mesh plot.

To construct this plot, make selections as indicated. Accept Transparent background color and black Mesh color.

Remember to review only one plot at a time or else plots will be overlaid; hide pressure and velocity plots prior to viewing the mesh plot. Review the mesh plot in Figure 14-22 and notice that the cavity where flow occurs is meshed and the surrounding walls are not. This is because heat flow through solids has not been requested in this exercise so there was no need to mesh the solid geometry.

Remember to hide CAD geometry using the **Display Model Geometry** command in the **Flow Simulation** Command Manager (Figure 14-13).

We now move to a **Conjugate** heat transfer problem where a fluid flow is accompanied by a heat flow. Open the assembly model HEATER shown in Figure 14-23. Our objective is to find heat power and thermal expansion in this water heater. Along the way we will have to analyze fluid flow to find convective boundary condition between flowing water and solid brass tube. The water heater is surrounded by air but since we won't model air flow, the convective boundary conditions on all of the outside faces must be given; we assume the convection coefficient is $50W/(m^2K)$ and the ambient temperature is 293.2K.

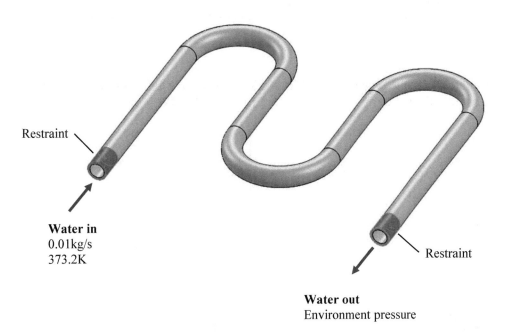

Restraint

Water in
0.01kg/s
373.2K

Restraint

Water out
Environment pressure

Figure 14-23: HEATER assembly model.

The restraints used in structural analysis are shown. Plugs on the inlet and the outlet are not shown.

The following parameters define the flow of water and the conjugate heat flow:

> The model is made watertight by plugging up the inlet and outlet
> Flow in: 0.01kg/s, temperature 373.2K
> Flow out: atmospheric pressure
> Convection coefficient on all outside faces: $50W/(m^2K)$
> Ambient temperature 293.2K
> Material: brass

The following parameters define the structural analysis:

Two ends marked by split lines are restrained; the restraints will be defined in such a way as to not interfere with thermal expansion

Case 1
Loads are temperatures imported from the **Thermal** study.
Branch 1 in Figure 14-1

Case 2
Loads are temperatures imported directly from **Flow Simulation**.
Branch 2 in Figure 14-1

Use the **Wizard** to create a **Flow Simulation** project, and call it *Project 02*. We show only the windows that are different from the ELBOW project.

Select Internal   Select Heat conduction in solids

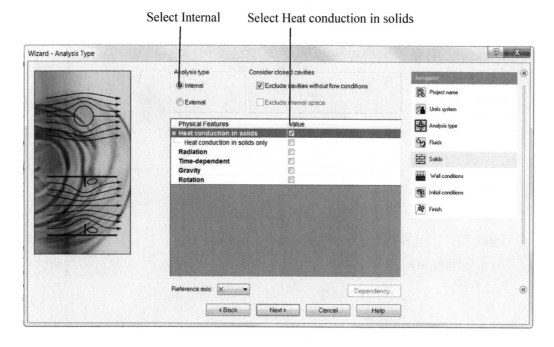

Figure 14-24: Analysis type.

Select Heat conduction in solids: this way heat transfer in the fluid and in the solid will be modeled.

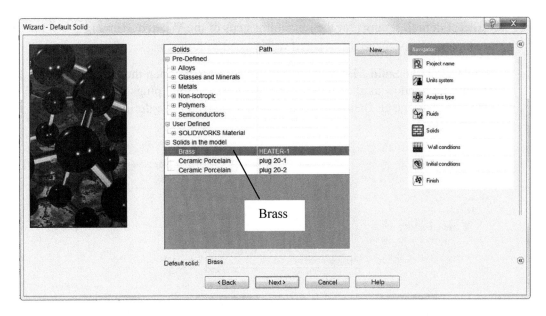

Figure 14-25: Default Solid.

Select Brass from the "Solids in the model" list.

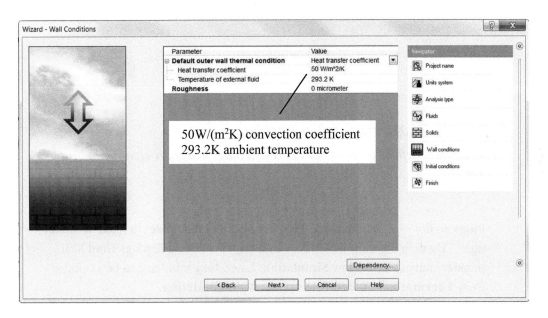

Figure 14-26: Wall Conditions.

Define a Heat transfer coefficient of 50W/(m²K) as the default convection coefficient on the outside walls. This "imposed" definition is required because fluid flow on the outside of the HEATER is not modeled.

Accept default settings for the remaining steps of the **Wizard**. Having completed the steps in the **Wizard** we proceed as follows:

Right-click the **Solid Materials** folder in *Project 02* to open the **Solid Material** definition window as shown in Figure 14-27. Select both plugs from the fly-out menu and assign an **Insulator** material from the list of pre-defined materials.

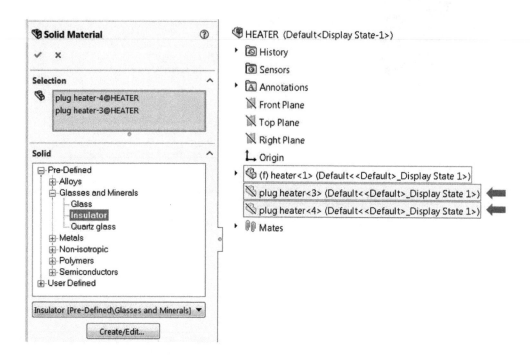

Figure 14-27: Material definition for plugs.

Material is defined as Insulator to prevent plugs from participating in heat transfer.

Plugs do not exist as real parts. They are added to the model to make it "water-tight." By defining their material as **Insulator**, we exclude plugs from heat transfer analysis with **Flow Simulation**. Later, they will have to be excluded from **Thermal** and **Structural** analyses with **Simulation**.

Define flow boundary conditions on the inlet as shown in Figure 14-28. Flow is 0.01kg/s; water is at boiling temperature of 373.2K.

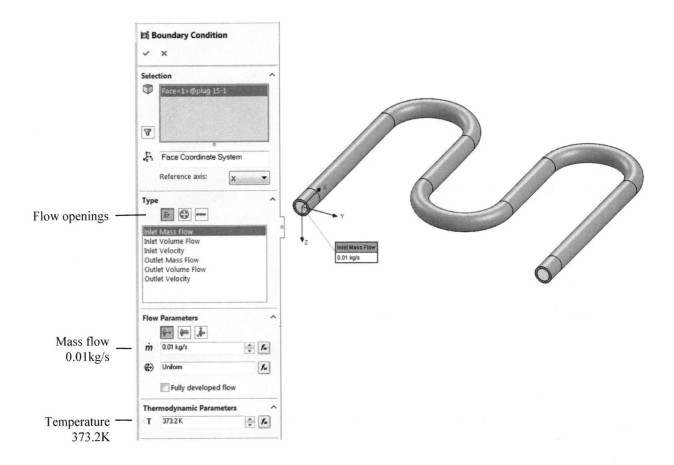

Flow openings

Mass flow
0.01kg/s

Temperature
373.2K

Figure 14-28: Flow boundary conditions in the water inlet. This is a slow flow of 0.01kg/min.

Plugs are shown in this illustration. Use "Select other" command to select the back face of the plug.

Define flow boundary conditions on the outlet as shown in Figure 14-29.

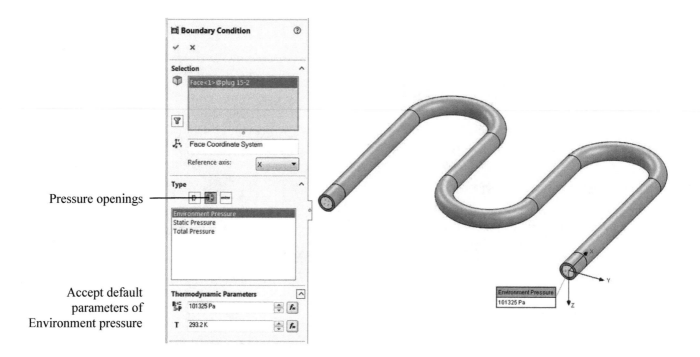

Pressure openings

Accept default parameters of Environment pressure

Figure 14-29: Flow boundary conditions on the outlet.

Accept default parameters of environment pressure because water exits to the environment.

The fluid flow is now fully defined.

Define the solution **Goals** as the average, the Maximum, the Average and the Minimum temperature in solids (Figure 14-30). These Global Goals are defined in addition to internal solver settings.

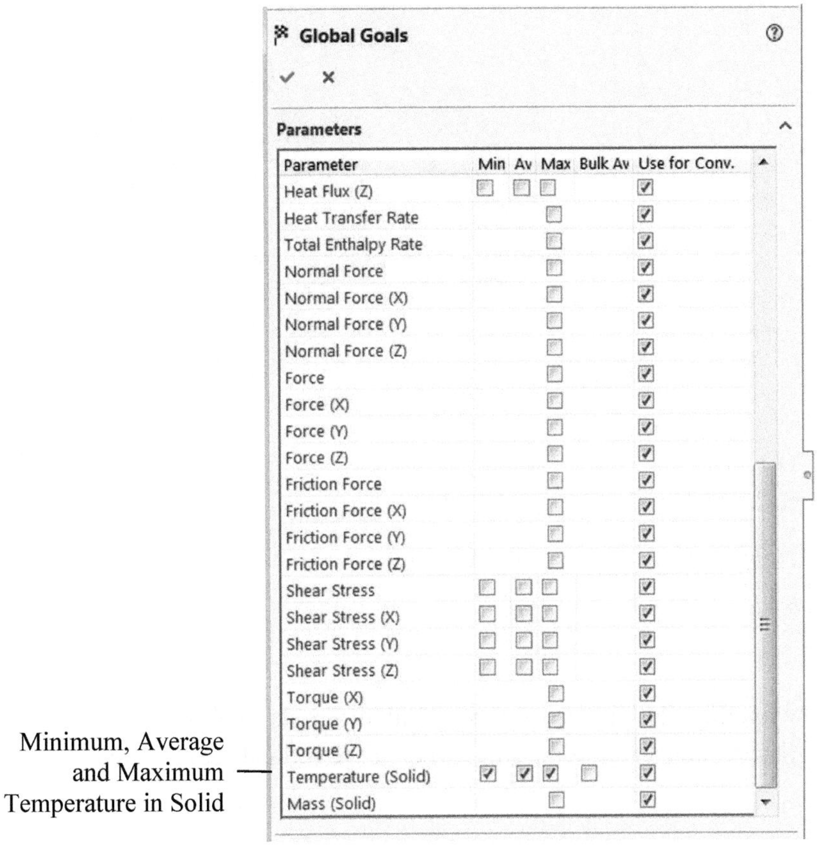

Minimum, Average and Maximum Temperature in Solid —

Figure 14-30: Definition of solution Goals.

Average, Minimum and Maximum Temperature in Solid are selected because we are particularly interested in thermal effects in the solid model. Fluid flow is modeled "out of necessity" to find convective boundary conditions on the inside faces of the water heater.

Set the level of initial mesh to 4 as shown in Figure 14-31.

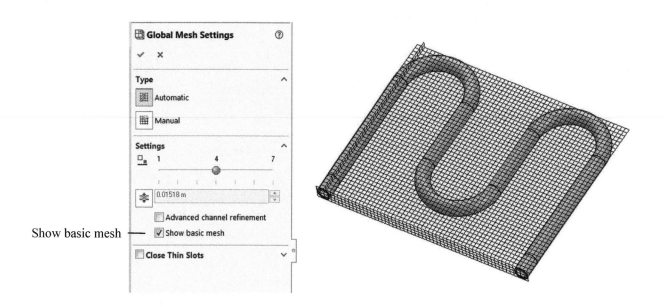

Show basic mesh

Figure 14-31: Global Mesh Settings set to level 4.

Select Show basic mesh to see this plot.

You may now run the solution; later repeat it with different levels of mesh refinements; observe longer solution times when more refined meshes are used.

Before proceeding with thermal and thermal stress analyses, review the results of the **Flow Simulation**, especially those that are related to subsequent analysis with **Simulation** (Figure 14-32).

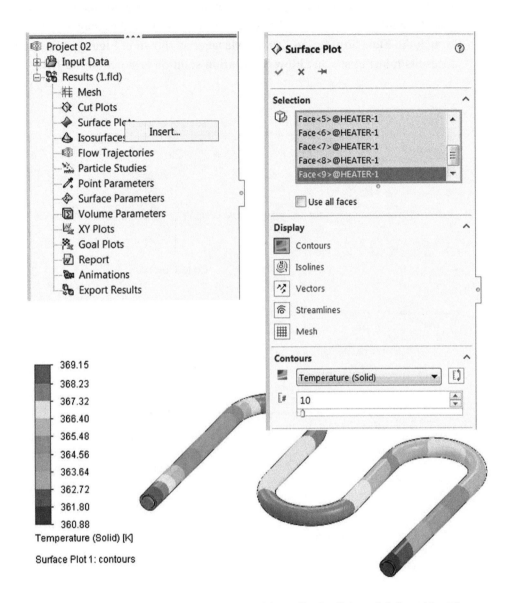

Temperature (Solid) [K]

Surface Plot 1: contours

Figure 14-32: Temperature of the outside wall of solid model found by Flow Simulation.

Inlet and outlet plugs as hidden.

Having solved the fluid flow problem, we have to transfer results from **Flow Simulation** to **Thermal** study and to **Static** study in **Simulation** but first we need to know where the results of **Flow Simulation** are stored.

Management of results in **Flow Simulation** is different from the system used in **Simulation**. Results files are named by numbers, which are not directly associated with the project name. They are located in folders with the same numerical names as file names. Results may be loaded or unloaded using the **Load/Unload** command in the **Flow Simulation** Command Manager window or through the **Flow Simulation** Feature Manager as shown in Figure 14-33. Look at Results folder to see that **Flow Simulation** solution is stored in file **1.fld**.

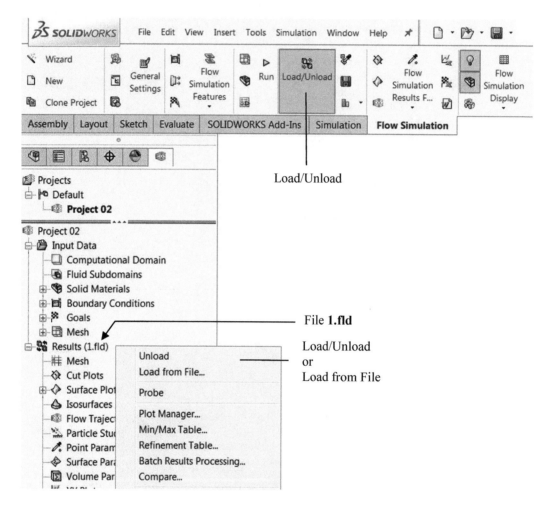

Figure 14-33: Load/Unload can be executed through the pop-up menu invoked by right-clicking on the Results folder or through the Flow Simulation Feature Manager window.

The load command will load default results associated with the project. You may also load other results selecting Load from File. Your *.fld file may have a different number assigned to it.

Refer to Figure 14-34 for information on how to find the **Flow Simulation** project results.

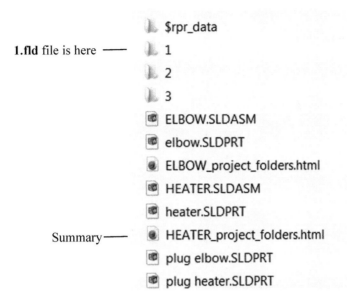

1.fld file is here ——

Summary——

Figure 14-34: Content of folder 14 Heat transfer with fluid flow.

Folders 1, 2, 3, ... are created by Flow Simulation.

Your result files may be numbered differently depending on the order in which you have completed the exercises. Summary of projects related to each model is stored in an *.html file. You may open them with a browser such as Firefox or Chrome.

Results of **Flow Simulation** must be transferred to **Simulation** as shown in Figure 14-35. Save the model before and after exporting the results.

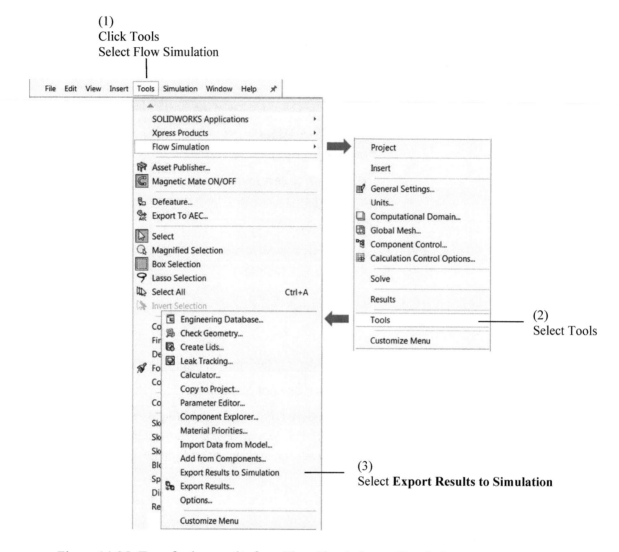

Figure 14-35: Transferring results from Flow Simulation to Simulation.

Each step indicated above opens a new window. The above windows have been modified to improve clarity of this illustration. Save the model before and after exporting results to Simulation.

Notice that when you move to **Simulation**, the model display and its visibility as well as display of **Flow Simulation** results plots are still affected by visibility settings in **Flow Simulation**. Learn how to move between windows of **SOLIDWORKS** Feature Manager, **Simulation** studies and **Flow Simulation** projects.

Hide the plugs from view (but do not suppress them) in **SOLIDWORKS** Feature manager and create a **Thermal** study *01 Thermal*.

We will now follow Branch 1 in Figure 14-1. In *01 Thermal* study properties select **Include fluid convection effects from SOLIDWORKS Flow Simulation** and select the correct **Flow Simulation** project file (Figure 14-36).

Select **Include fluid convection effects**

Select the **Flow Simulation** results file.
Here the results file is 1.**fld** as shown in Figure 14-34

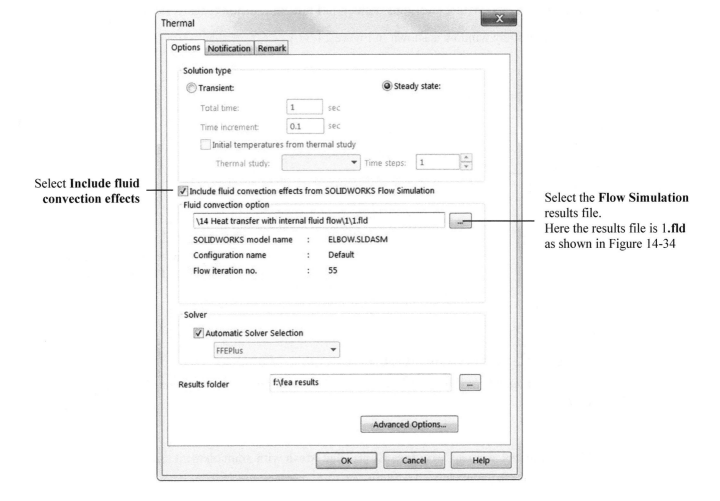

Figure 14-36: Importing convection coefficients from Flow Simulation results to Thermal study in Simulation.

If you have more than one Flow Simulation project, make sure to select the correct Flow Simulation results file.

Importing **Flow Simulation** results as shown in Figure 14-36, only brings into the **Thermal** study the convective boundary conditions that have been calculated in **Flow Simulation**. Convection coefficients and ambient temperatures that have been "imposed" in a project definition (Figure 14-26) must be defined in the **Thermal** study.

Exclude both plugs from the analysis and define a convection coefficient of $50W/(m^2K)$ with an ambient temperature of 293.2K on all external faces. This corresponds to the **Wall Conditions** defined in **Flow Simulation** (Figure 14-26).

01 Thermal study window is shown in Figure 14-37.

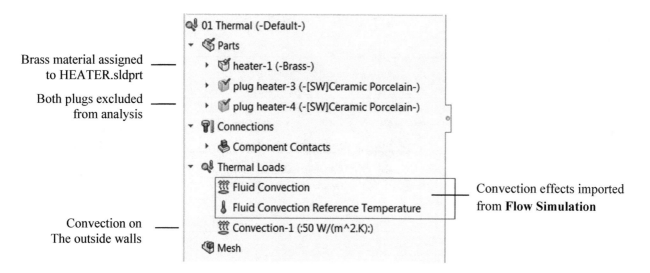

Brass material assigned to HEATER.sldprt

Both plugs excluded from analysis

Convection effects imported from **Flow Simulation**

Convection on The outside walls

Figure 14-37: Thermal 01 study ready to run.

Plugs are hidden in SOLIDWORKS and excluded from analysis in 01 Thermal study.

Mesh the model using a **Curvature Controlled Mesh** with 3mm element size; leave other mesh parameters on defaults.

Thermal conditions imported from **Flow Simulation** can be reviewed as shown in Figure 14-38.

Right-click "Fluid Convection" imported from **Flow Simulation**

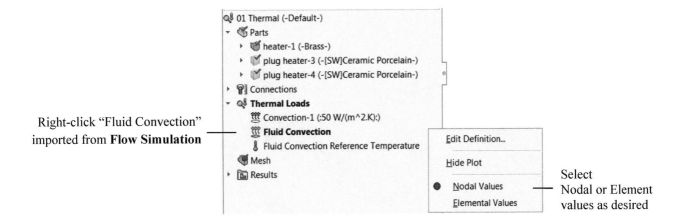

Select Nodal or Element values as desired

<u>Figure 14-38: Displaying convection coefficients imported from Flow Simulation.</u>

Follow the same steps to display the Fluid Convection Reference Temperature plot. Displaying Fluid Convection and Fluid Convection Reference Temperature will work only if the model has already been meshed.

In this project, the imported Fluid Convection coefficients and Fluid Convection Reference Temperatures are defined on the inside faces of the manifold. Since section plots are not available for these types of plots, the review of plots is limited to openings on the side of inlets and outlet. Because of these limitations the plots are not shown here. Review Fluid Convection plot and Fluid Convection Reference Temperature plot in colors.

Run *01 Thermal* study and display a **Temperature** plot (Figure 14-39).

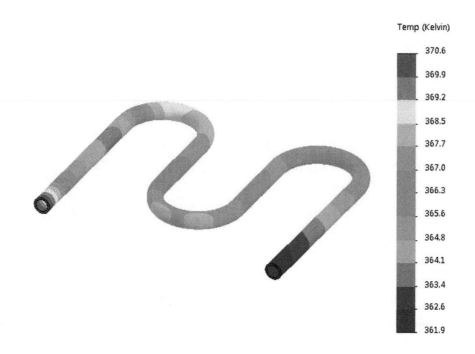

Figure 14-39: Temperature results from Thermal analysis.

This plot is very similar to the temperature in the solids plot in Flow Simulation shown in Figure 14-32.

The difference between temperature results for solids reported by **Flow Simulation** and **Thermal** analysis in **Simulation** is caused by the error of translation of fluid convection effects from **Flow Simulation** to **Simulation** and by discretization errors in both applications.

You may now use these **Thermal** analysis results to analyze heat flux, heat power etc. This would not be possible within **Flow Simulation, which** is not intended for the analysis of solids.

Heat Power results are shown in Figure 14-40.

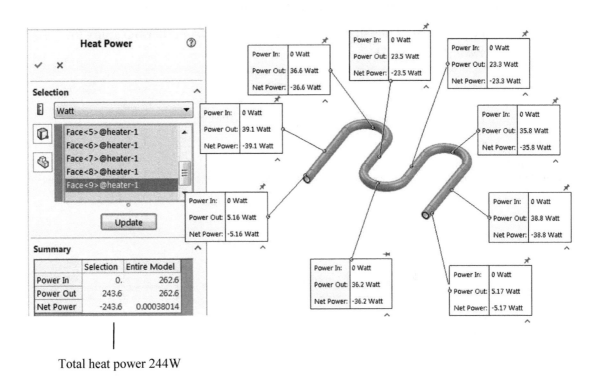

Total heat power 244W

Figure 14-40: Heat Power results from Thermal analysis.

Callouts may be pinned for easier review of results.

We will now perform an analysis of thermal expansion. We may proceed in two ways: temperature results may be imported either from **Thermal** study (branch 1 in Figure 14-1) or from **Flow Simulation** (branch 2 in Figure 14-1). We start with the first method. Create a static study called *02 Thermal Expansion*.

In the properties of this study specify **Temperature from thermal study** *01 Thermal* (Figure 14-41).

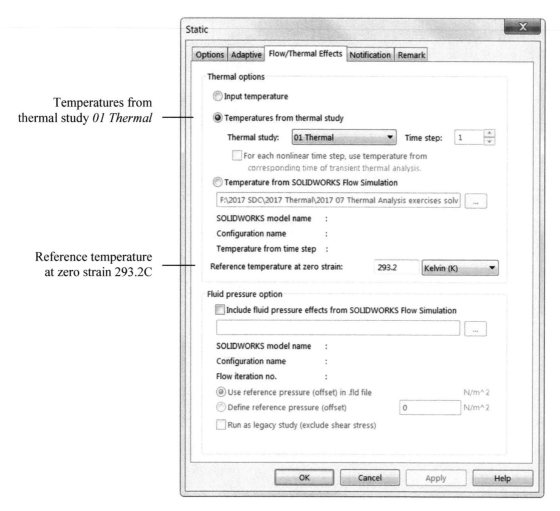

Figure 14-41: Properties of *02 Thermal Expansion* study.

The reference temperature at zero strain corresponds to the temperature of the external fluid defined in Figure 14-26.

Verify that Brass material is assigned to HEATER, exclude plugs from analysis and copy the mesh from *01 Thermal* study.

As in any structural analysis, the model must have proper restraints that eliminate Rigid Body Motions. We will define restraints that fully restrain the model but avoid interfering with thermal expansion (Figure 14-42).

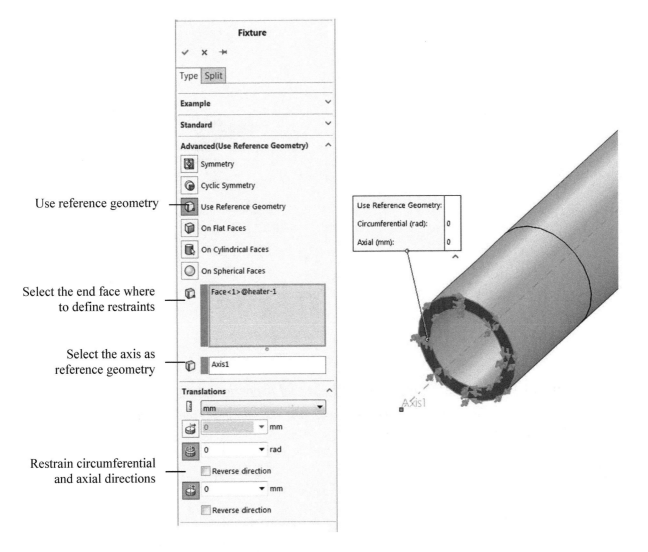

Figure 14-42: Restraints in *02 Thermal Expansion* study defined on the end face. This restraint may be defined on either one of the two end faces.

This restraint eliminates all rigid body motions but does not interfere with thermal expansion.

Run the solution and review displacement results (Figure 14-43).

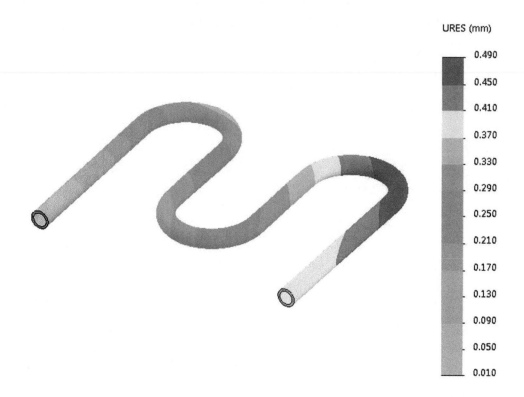

Figure 14-43: Displacement results from *02 Thermal Expansion study.*

Displacements are measured from the end when the restraint has been defined (Figure 14-42).

We conclude the HEATER exercise by demonstrating the analysis along Branch 2 (Figure 14-1). Copy study *02 Thermal Expansion* into *03 Thermal Expansion*. Modify the properties of *03 Thermal Expansion* study as shown in Figure 14-44.

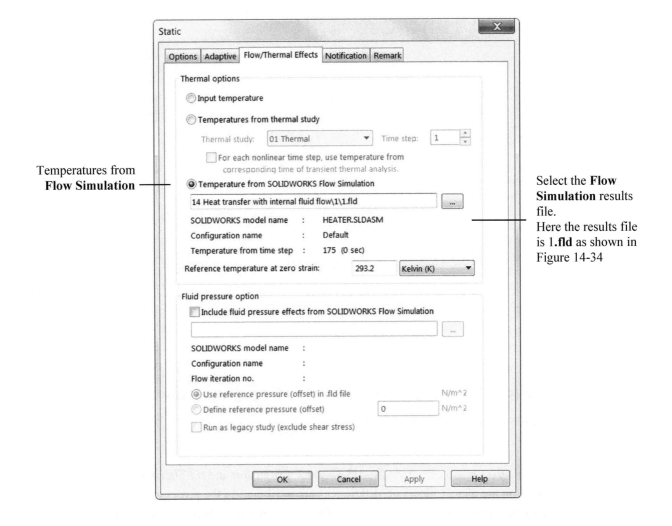

Temperatures from **Flow Simulation**

Select the **Flow Simulation** results file.
Here the results file is **1.fld** as shown in Figure 14-34

Figure 14-44: Properties of *03 Thermal Expansion* study.

Temperatures are imported directly from the Flow Simulation project.

The change in the study property is the only modification required in study *03 Thermal Expansion*.

Run the solution and display stress results (Figure 14-45).

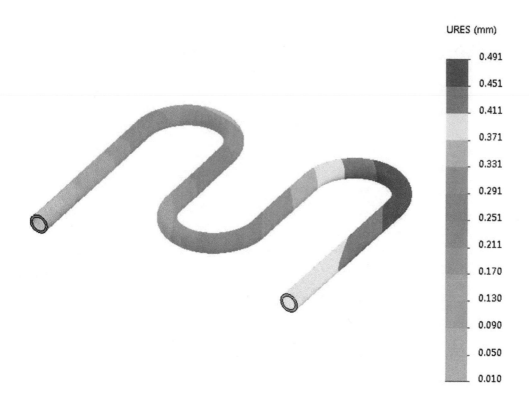

URES (mm)

0.491
0.451
0.411
0.371
0.331
0.291
0.251
0.211
0.170
0.130
0.090
0.050
0.010

Figure 14-45: Displacement results from *03 Thermal Expansion* study.

Compare these results to stress results shown in Figure 14-44.

Thermal stress results from studies *02 Thermal Expansion* and *03 Thermal Expansion* are almost identical; differences are attributed to the errors in transfer of convective boundary conditions from **Flow Simulation** to **Simulation** and to errors of discretization error.

In this introductory exercise, we have demonstrated two ways of interfacing between **Flow Simulation**, **Thermal,** and **Static** (Figure 14-1). If thermal analysis in solids is required, you have to follow Branch 1. If thermal stresses only are required, then direct interfacing along Branch 2 is faster.

You may want to extend this exercise to include pressure effects from **Flow Simulation** in a **Static** analysis. To obtain a meaningfully high stress, increase water flow in the heater.

Summary of studies completed

Model	Configuration	Study or Project Name	Study Type
ELBOW.sldasm	*Default*	*Project 01*	Flow Simulation
HEATER.sldasm	*Default*	*Project 02*	Flow Simulation
		01 Thermal	Thermal
		02 Thermal Expansion	Static
		03 Thermal Expansion	Static

Figure 14-46: Names and types of studies completed in this chapter.

Notes:

15: Heat transfer with external fluid flow

Topics covered

- Using Flow Simulation for finding convection coefficients in external fluid flow
- Interfacing between Flow Simulation and Thermal analysis

Project description

Figure 15-1 explains the differences between an internal and an external fluid flow.

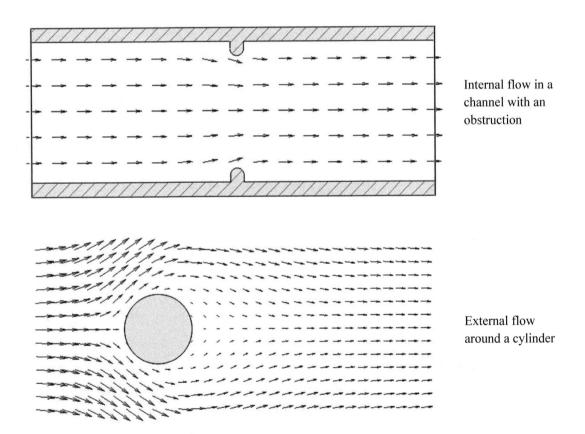

Internal flow in a channel with an obstruction

External flow around a cylinder

Figure 15-1: Internal and external fluid flow.

Velocity vectors are shown in both illustrations. Channel is shown in a cross-section.

The illustrations in Figure 15-1 were prepared using models INTERNAL FLOW and EXTERNAL FLOW. These models come with **Flow Simulation** projects defined so it is easy to replicate these plots. The INTERNAL FLOW model uses automatically created plugs, which have not been used in the previous chapter.

To study heat transfer in the presence of external fluid flow we will analyze a brass sphere submerged in slow flowing water. The heat power of the sphere is 1000W. The temperature of the cooling water is 293.2K; water is under atmospheric pressure and the flow velocity is 0.1m/s.

Open part model BALL15 which is shown in Figure 15-2. The same figure also shows the **Computational Domain** automatically defined by the **Flow Simulation** wizard when external flow is specified.

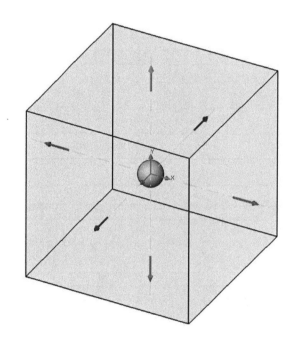

Figure 15-2: Model BALL15 and the Computational Domain.

The Computational Domain extends to the region where the external flow is no longer disturbed by the immersed object.

Use the **Wizard** to define a **Flow Simulation** project. Figures 15-3 through 15-9 show windows corresponding to steps performed using the **Wizard**. Compare them with the **Wizard** steps from the previous chapter where we analyzed internal flows.

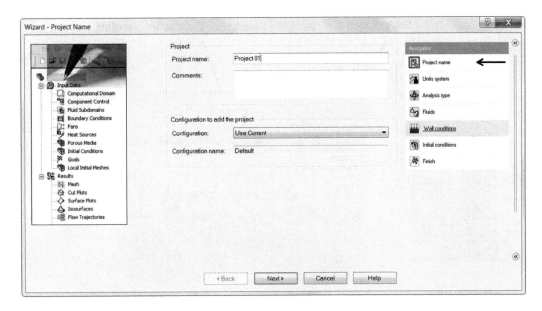

Figure 15-3: Step 1: Project name.

You may move through Wizard windows using the Next and Back buttons or by selecting steps from the Navigator menu on the right.

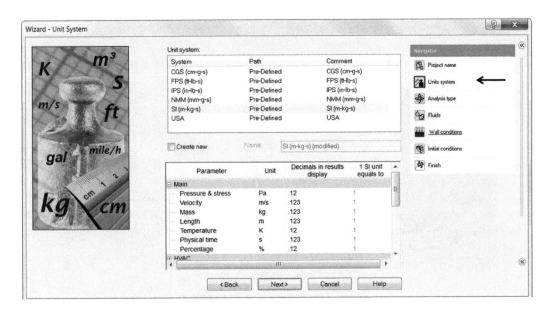

Figure 15-4: Step 2: Unit system.

Accept the default SI system of units.

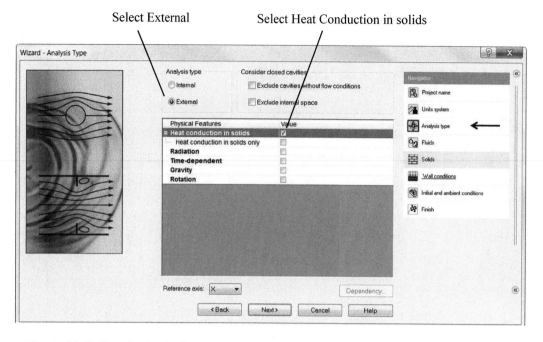

Figure 15-5: Step 3: Analysis type.

Select the External type of analysis; select Heat conduction in solids.

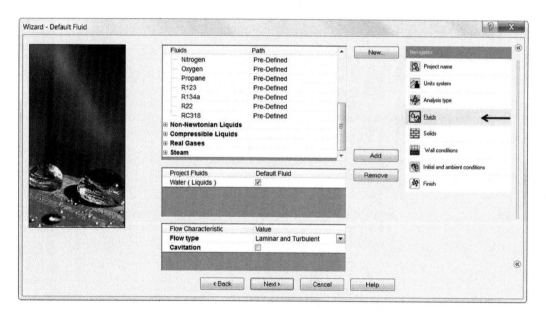

Figure 15-6: Step 4: Default fluid.

Select water from the list of pre-defined fluids.

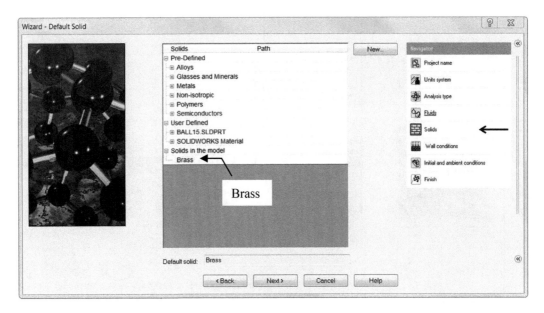

Figure 15-7: Step 5: Default Solid.

Select Brass from "Solids in the model."

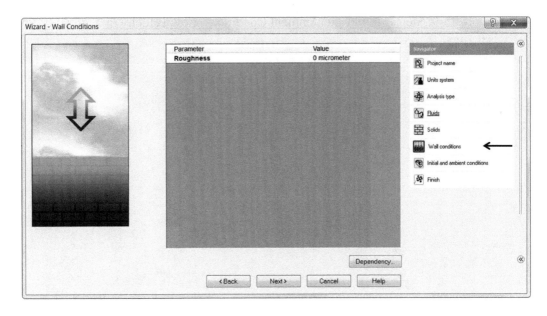

Figure 15-8: Step 6: Wall Conditions.

Do not change anything here; the ball surface is assumed perfectly smooth.

Flow velocity in X direction is 0.1m/s

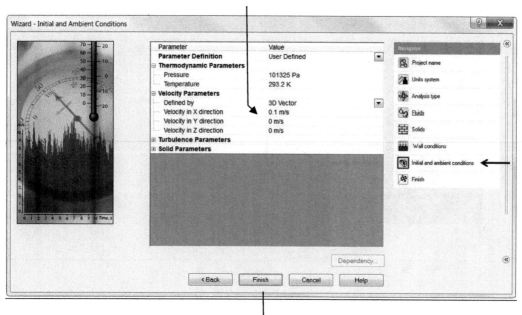

Click Finish to exit Wizard

Figure 15-9: Step 7: Initial and Ambient Conditions; flow velocity in X direction is 0.1m/s.

This defines the velocity of the external flow.

Set **Level of initial mesh** to 6 and **Refinement level** to 2 (Figure 15-10).

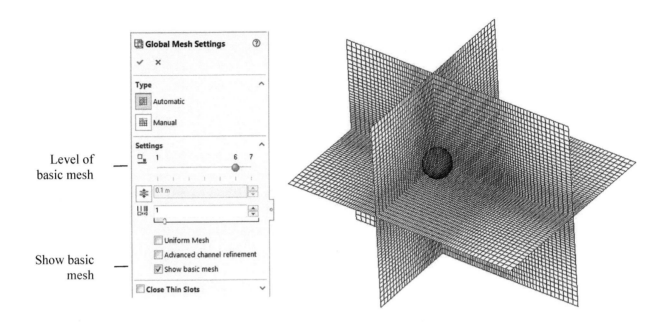

Level of
basic mesh

Show basic
mesh

Figure 15-10: Global Mesh Settings.

We select resolution level 6 for precise transfer of convective boundary conditions to Simulation. The basic mesh is shown here on three cross-section planes.

Using this high level of the initial mesh resolution (level 6), we do not have to use local mesh refinement but the project will take a long time to solve.

Follow the steps shown in Figure 15-11 to define the 1000W heat power generated in the entire volume of the model.

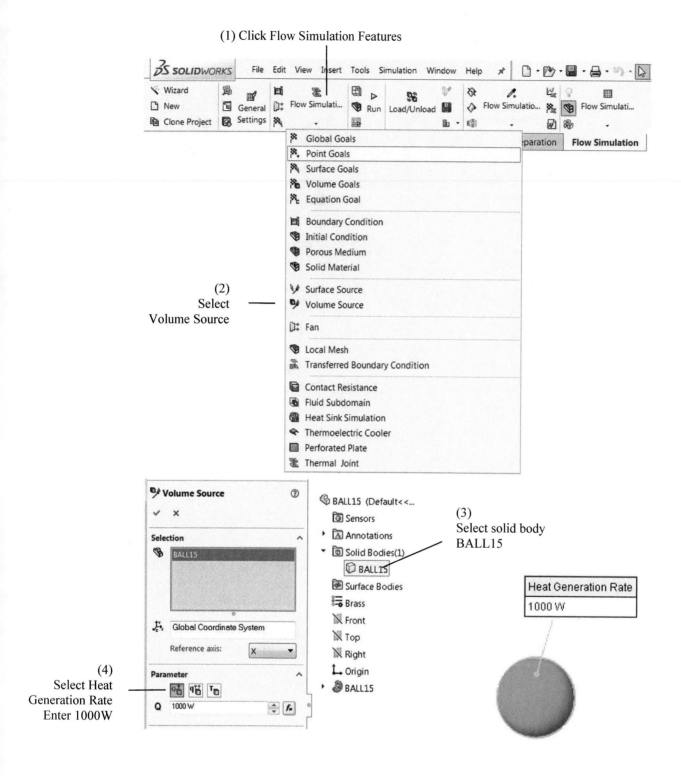

(1) Click Flow Simulation Features

(2) Select Volume Source

(3) Select solid body BALL15

(4) Select Heat Generation Rate Enter 1000W

Figure 15-11: Definition of heat power.

Heat Generation Rate applies heat power to the entire volume of the selected solid body.

The objective of this analysis is finding the temperature of the ball cooled by the flow of water. Therefore, create Goals as shown in Figure 15-12.

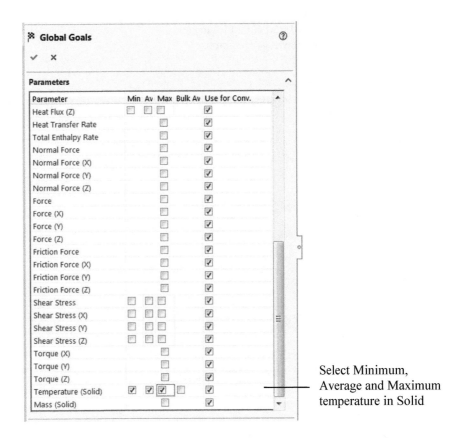

Figure 15-12: Definition of analysis goals.

The minimum, the average and the maximum temperature will be used in the convergence check along with pre-defined internal checks in solver.

Run the solution and observe the convergence process available in the Solver window as shown in Figure 15-13.

Click Insert Goal Plot to see live convergence graphs

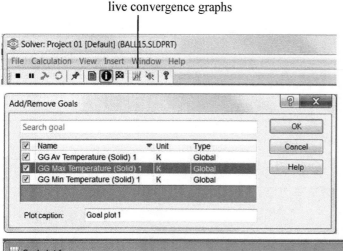

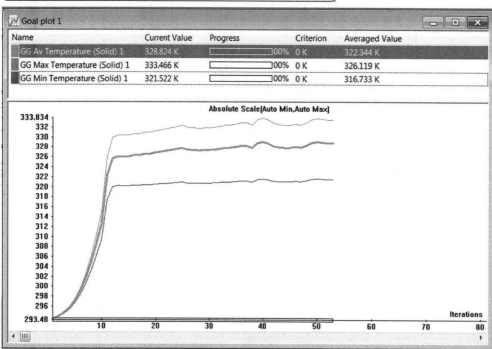

Figure 15-13: Monitoring solution progress in the Solver window.

Review other options available in the Solver window.

Be prepared for a long solution time.

Analysis of water flow is not the main objective of this analysis; still you may want to look at some results pertaining to flow. For example, create a **Cut Plot** of velocity as shown in Figure 15-14.

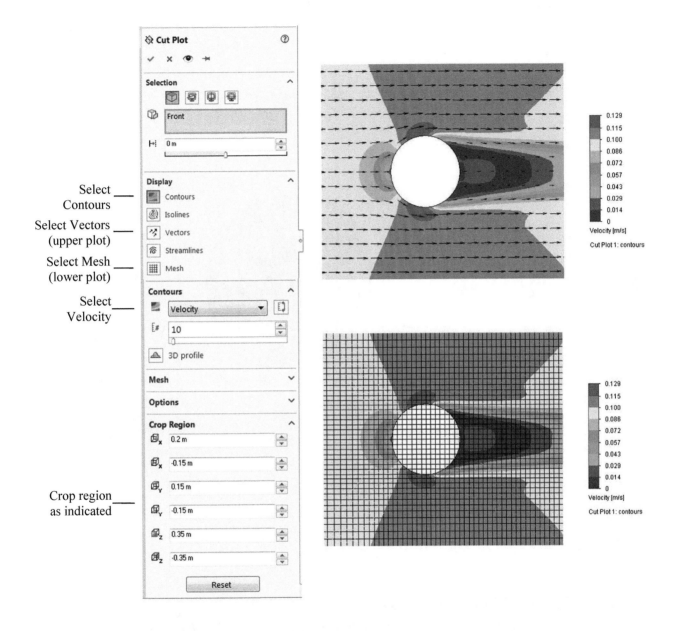

Select Contours

Select Vectors (upper plot)

Select Mesh (lower plot)

Select Velocity

Crop region as indicated

Figure 15-14: Velocity cut plot definition.

Contour plot of velocity has overlaid vectors (top) and overlaid mesh (bottom).

Define a **Cut Plot** showing temperature in the solid as shown in Figure 15-15.

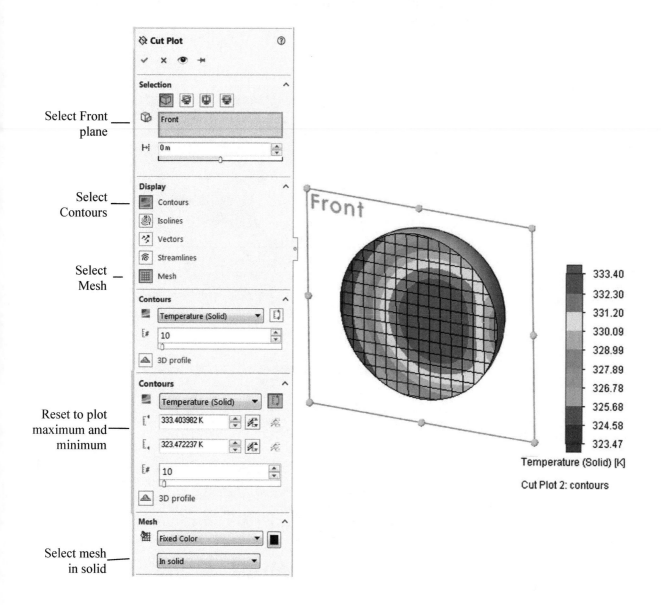

Select Front plane

Select Contours

Select Mesh

Reset to plot maximum and minimum

Select mesh in solid

Figure 15-15: Temperature in the cross section of the ball.

The mesh in the solid includes solid cells and partial cells. In this plot, the CAD model is not hidden but it is shown in a cross-section by Front plane.

Hide all **Flow Simulation** plots and display the model geometry.

Export the results of **Flow Simulation** to **Simulation** as explained in Figure 14-35. Remember to save the model before and after exporting the results.

Create **Thermal** study *Ball Thermal* in **Simulation**; in **Thermal** study properties created, select **Include fluid convection effects from SOLIDWORKS Flow Simulation**. Select the correct *.fld file as shown in Figures 14-33 and 14-34.

Define a **Heat Power** of 1000W in the volume and mesh the model with the default mesh.

All thermal conditions necessary to run the study have now been set-up. You may review the convective boundary conditions imported from **Flow Simulation** as shown in Figure 15-16. Model must be meshed in Thermal study before this plot can be shown.

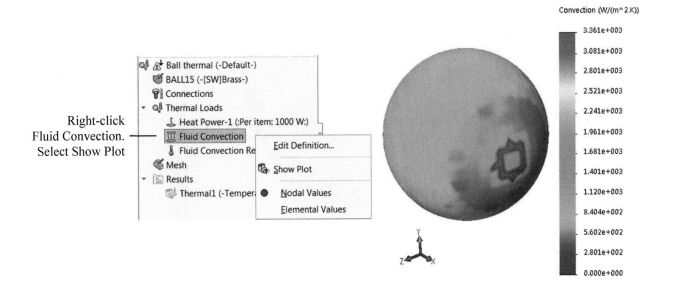

Figure 15-16: Convection boundary conditions imported from Flow Simulation.

The irregular shape of the fringes and concentration is a result of translation error from Flow Simulation to the Thermal study in Simulation and of discretization errors in both applications. Repeat the Ball Thermal study with a more refined mesh.

In continuation of the plot review in Figure 15-16, display element values of the **Fluid Convection** plot. Once this is done, review the plot of **Fluid Convection Reference Temperature**.

Run the solution of the **Thermal** study and display the temperature plot (Figure 15-17).

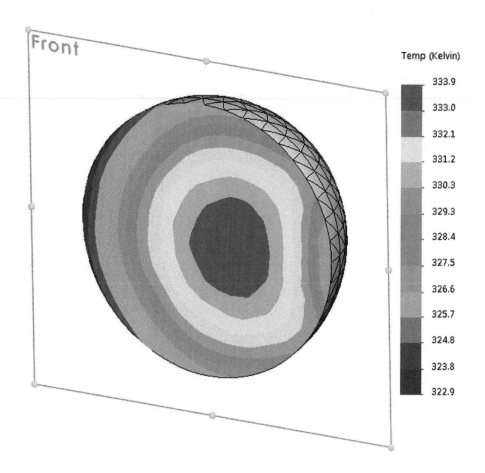

Figure 15-17: Temperature in the cross section of the ball.

Section clipping makes this plot directly comparable to the plot in Figure 15-15. Contours are not plotted on the uncut portion of the model, but mesh is shown there.

Plot in Figure 15-17 does not show mesh on the cross section. As opposed to the finite volume mesh (Figure 15-15) which is aligned with the Front plane, the finite element mesh is not aligned with the Front plane. You may use Mesh Sectioning plot to show the finite element mesh across the cross-section.

Having obtained results of **Thermal** analysis, you may wish to proceed to a detailed analysis of thermal conditions in the ball.

This exercise offers a good opportunity to study the effect of grid density in **Flow Simulation** on the accuracy of flow simulation and the accuracy of transfer of convective boundary conditions from Flow **Simulation** to **Thermal** analysis in **Simulation**, but this is out of the scope of this book.

Summary of studies completed

Model	Configuration	Study or Project Name	Study Type
INTERNAL FLOW.sldprt	*Default*	*Internal flow*	Flow Simulation
EXTERNAL FLOW.sldprt	*Default*	*External flow*	Flow Simulation
BALL15.sldprt	*Default*	*Project 01*	Flow Simulation
		Ball thermal	Thermal

Figure 15-18: Names and types of studies completed in this chapter.

Notes:

16: Radiative Heat Transfer

Topics covered

- ❑ Radiative heat transfer problem analyzed with Thermal Study in Simulation
- ❑ Radiative heat transfer problem analyzed with Flow Simulation

Project description

A graphite sphere produces a 2000W heat power in its volume. The heat is radiated out of the ball into space; some of the heat is reflected by the steel reflector. The ball and reflector are in a vacuum; therefore, there is no convection involved in the heat transfer. This presents a **Surface to surface** radiation problem with an **Open system** and is easy to solve with **Thermal** study in **Simulation**. What makes this problem interesting is that it can also be solved with **Flow Simulation**. Solving the same problem with **Finite Volume Method (FV)** implemented in **Flow Simulation** and with **Finite Element Method (FEM or FEA)** implemented in **Thermal** study in **Simulation** offers a comparison between these two numerical methods as well as comparison of how they have been implemented in **SOLIDWORKS**.

Open the RADIATIVE HEAT TRANSFER assembly shown in Figure 16-1.

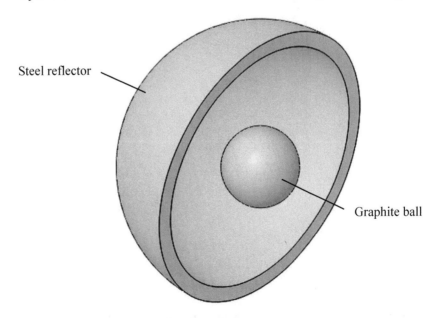

Steel reflector

Graphite ball

Figure 16-1: The graphite ball is partially surrounded by a steel reflector.

The assembly is in a vacuum; therefore, convective heat transfer does not take place.

Procedure

We will first solve this problem with **Flow Simulation**; then we will solve it the "standard way" with a **Thermal** study, matching as closely as possible all modeling assumptions introduced in the **Flow Simulation** study. Results of both studies will be close but not identical.

Use the **Flow Simulation Wizard** to set up the project *FVM 01* as shown in Figures 16-2 through 16-7.

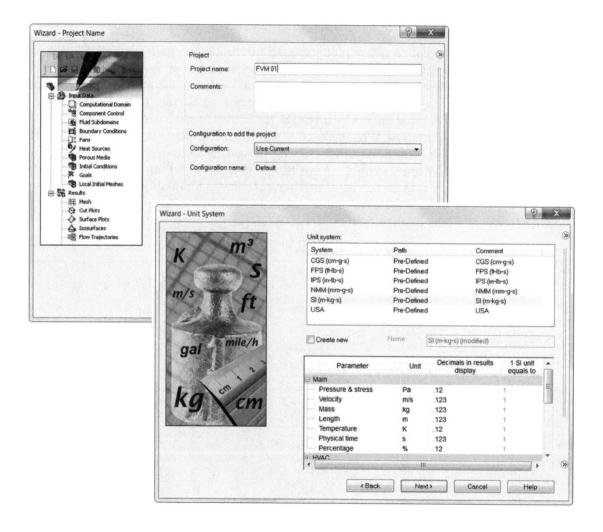

Figure 16-2: Steps 1 and 2: Project Name and Unit System.

These windows are self-explanatory. Navigator window is not shown.

External analysis

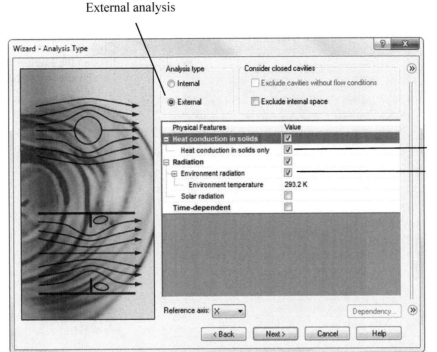

Heat conduction in solids and only in solids

Radiative heat transfer to be considered with an Environment temperature (ambient) of 293.2K

Figure 16-3: Step 3: Analysis Type.

Make selections as shown above to activate radiative heat transfer.

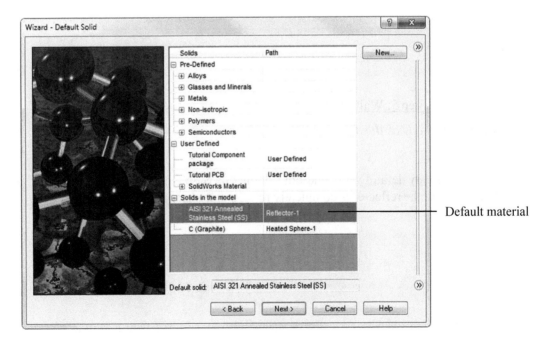

Default material

Figure 16-4: Step 4: Default solid.

Reflector material AISI 321 is the default material.

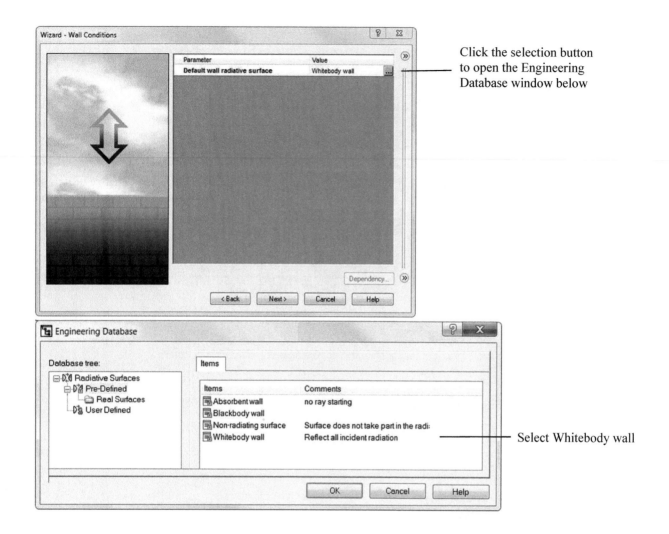

Click the selection button to open the Engineering Database window below

Select Whitebody wall

<u>Figure 16-5: Step 5: Wall conditions.</u>

Select Whitebody as the default wall condition.

The **Whitebody** default wall conditions defined in Step 5 means that all radiative heat reaching the reflector is completely reflected; no heat is absorbed by the reflector.

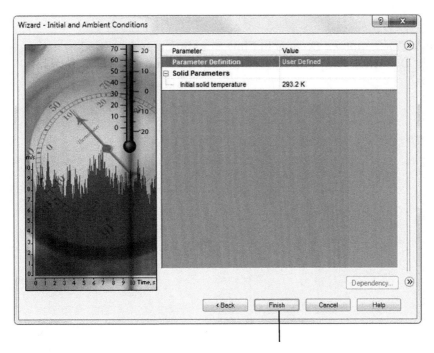

Click Finish to exit Wizard

Figure 16-6: Step 6: Initial and Ambient Conditions.

Accept the default Initial solid temperature of 293.2K.

Right-click *Computational Domain* in **Flow Simulation** project window and select **Edit Definition** to review the **Computational Domain** as explained in Figure 16-7.

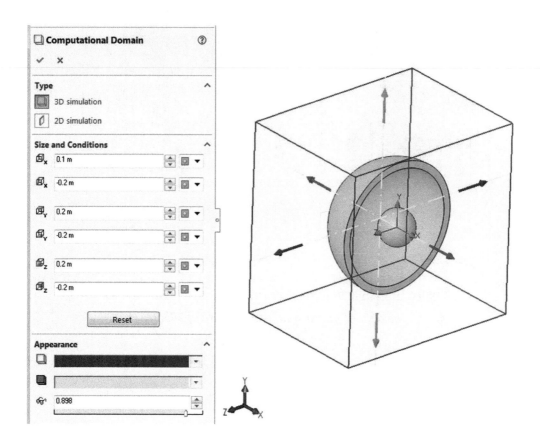

Figure 16-7: The automatically created Computational Domain can be modified using the Computational Domain window.

Appearance may be changed if desired.

Modification of the *Computational Domain* is not required in this exercise.

Define Global Mesh Settings as shown in Figure 16-8.

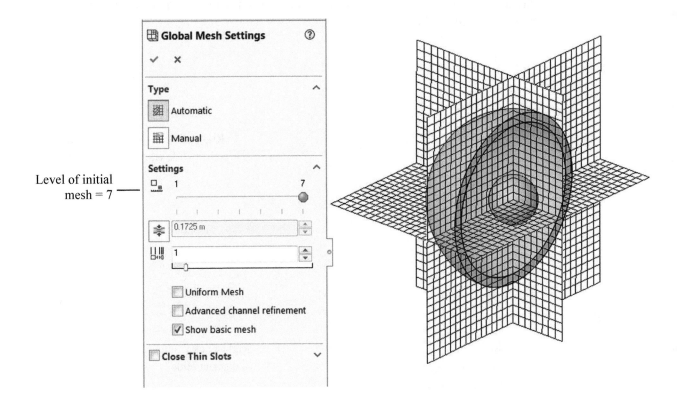

Figure 16-8: Global Mesh Settings.

Select Level of initial mesh 7. Mesh is shown on cross sections made by three orthogonal planes.

Having completed the initial steps using the project **Wizard**, modify the model display as explained in Figure 16-9.

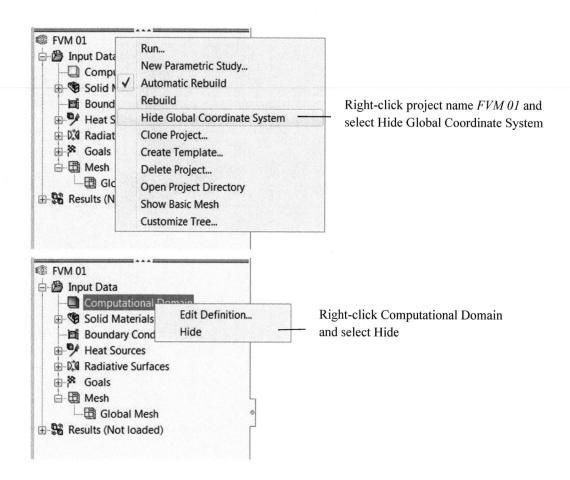

Right-click project name *FVM 01* and select Hide Global Coordinate System

Right-click Computational Domain and select Hide

Figure 16-9: Hide the Global Coordinate System and the Computational Domain.

These are "cosmetic" changes to improve the clarity of plots.

Define material of the graphite ball as shown in Figure 16-10.

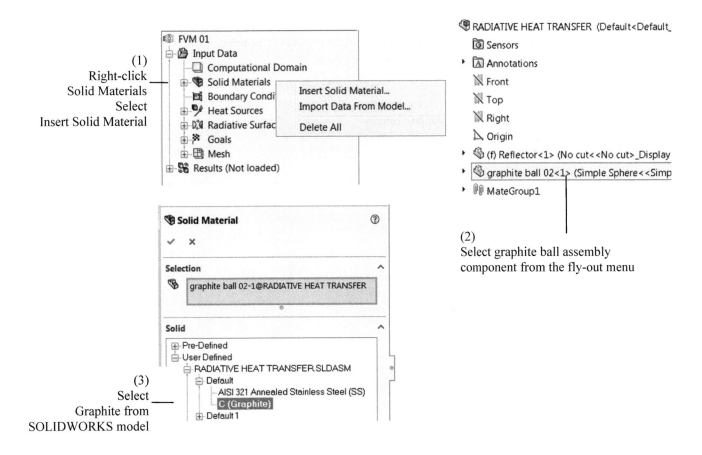

(1)
Right-click
Solid Materials
Select
Insert Solid Material

(2)
Select graphite ball assembly
component from the fly-out menu

(3)
Select
Graphite from
SOLIDWORKS model

Figure 16-10: Material assignment needs to be done for the graphite ball.

Assignment of the reflector material is not required because AISI321 has already been assigned as the default material in step 4 of the Wizard (Figure 16-4).

Now we will define heat power and radiative properties of the face of the graphite ball.

Follow the steps in Figure 16-11 to define a **Volume Source**. Follow the steps in Figure 16-12 to define a **Radiative Surface**.

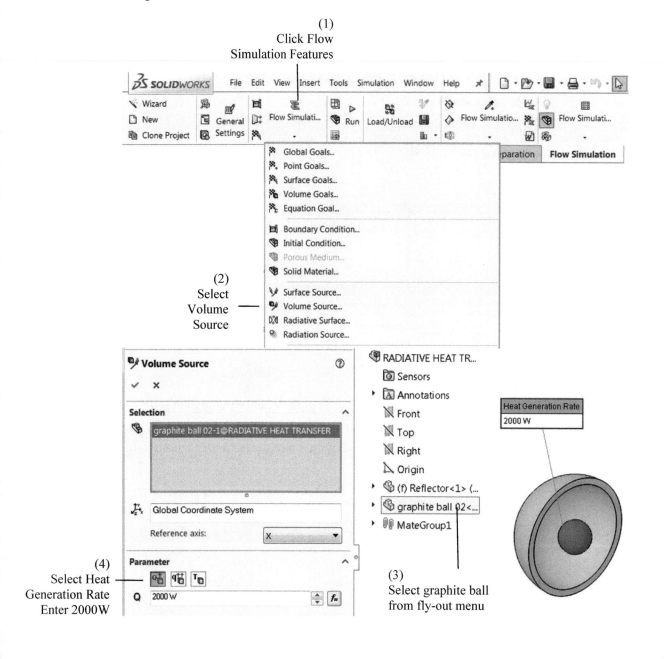

Figure 16-11: Definition of a Volume Source.

2000W of heat is generated in the volume of the graphite ball.

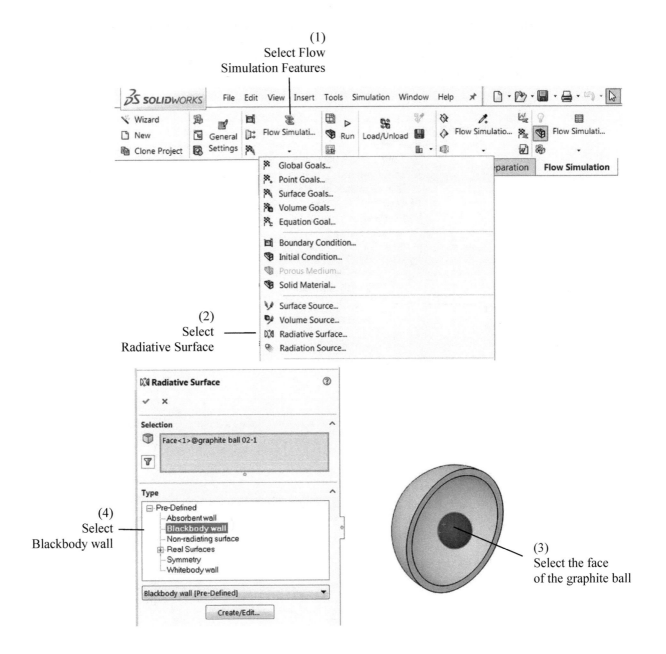

<u>Figure 16-12: Definition of a Radiative surface.</u>

A Blackbody radiative surface is the most efficient in radiating out and absorbing heat in the form of radiation.

The graphite ball radiates out heat. Some of that heat bounces off the reflector and returns to the graphite ball. Notice that we do not have to define **Radiative Surface** properties for the reflector. This has been already done in step 5 of the **Wizard** (Figure 16-5) where **Whitebody** radiative properties have been applied by defaults to all faces in the model. Those default **Whitebody** conditions have been overwritten on the face of graphite ball, which now has the properties of a **Blackbody**.

The temperature of the reflector will stay at 293.3K as defined in the **Wizard** (Figure 16-6) because the reflector has all **Whitebody** surfaces so no heat will enter it. We are interested in the temperature of the ball; therefore, define **Volume Goals** as shown in Figure 16-13.

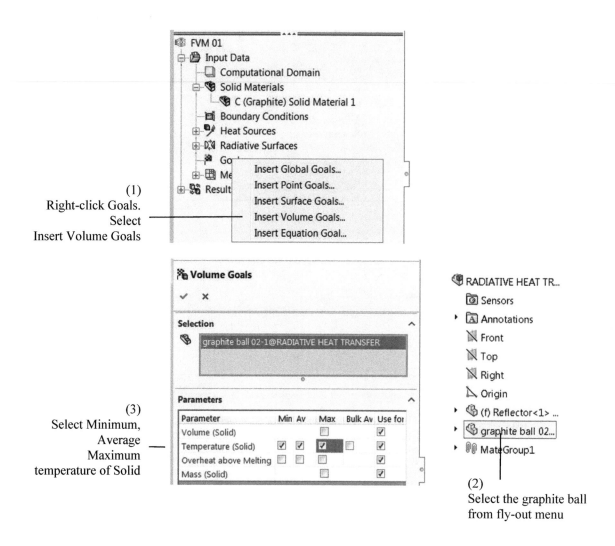

(1)
Right-click Goals.
Select
Insert Volume Goals

(3)
Select Minimum,
Average
Maximum
temperature of Solid

(2)
Select the graphite ball
from fly-out menu

Figure 16-13: Definition of Goals.

Minimum, averaged and maximum temperature in solids will be used for checking solution convergence.

Once the solution completes, create a **Cut Plot** of temperature as shown in Figure 16-14.

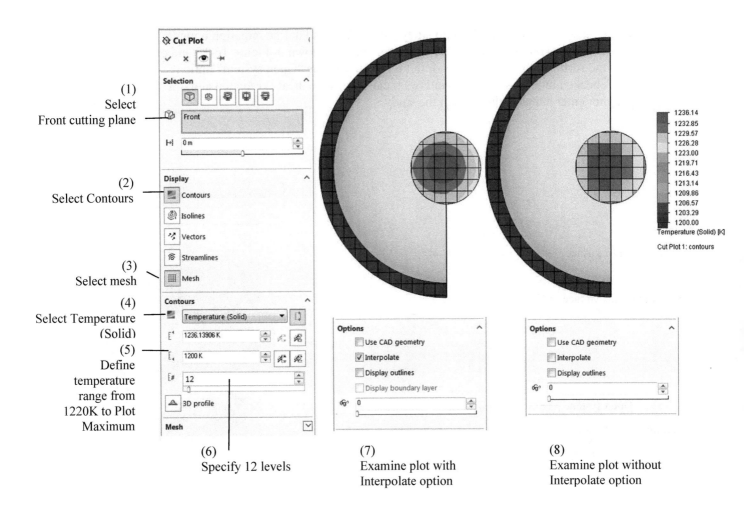

(1)
Select
Front cutting plane

(2)
Select Contours

(3)
Select mesh

(4)
Select Temperature
(Solid)

(5)
Define
temperature
range from
1220K to Plot
Maximum

(6)
Specify 12 levels

(7)
Examine plot with
Interpolate option

(8)
Examine plot without
Interpolate option

Figure 16-14: Temperature results.

We specify 12 levels in the color legend for consistency with Thermal analysis in Simulation, which uses 12 colors by default. The color legend may be moved around with the mouse.

Compare between interpolated and not interpolated plots.

Try to reset the temperature plot to show global minimum and notice that the reflector temperature stays at 293.2K as defined in **Default Wall Conditions**. This is because of the **Whitewall** default surface condition. The reflector does not absorb any heat.

We will now repeat the analysis using a **Thermal** study in **SOLIDWORKS Simulation**. We will match analysis assumptions made in **Flow Simulation** as closely as possible.

Create a steady state **Thermal** study called *FEM* and define radiation boundary conditions on the concave face of the reflector as shown in Figure 16-15 and on the surface of the ball as shown in Figure 16-16. The "Open system" option checked in both Radiation windows means that some heat is radiated and does not enter either of the parts in the study.

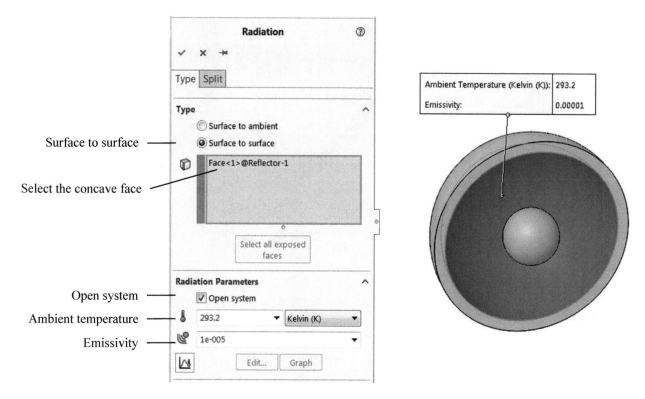

Figure 16-15: Radiation boundary conditions on the face of reflector.

Emissivity equal to zero defines White body conditions. Here we enter a very low emissivity 0.00001 because Thermal analysis will not run with zero emissivity.

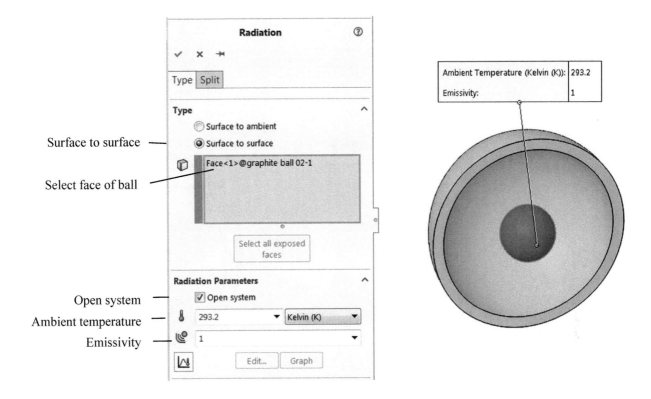

Figure 16-16: Radiation boundary conditions on the surface of the ball.

An Emissivity = 1 defines Blackbody conditions.

Define **Heat Power** as shown in Figure 16-17.

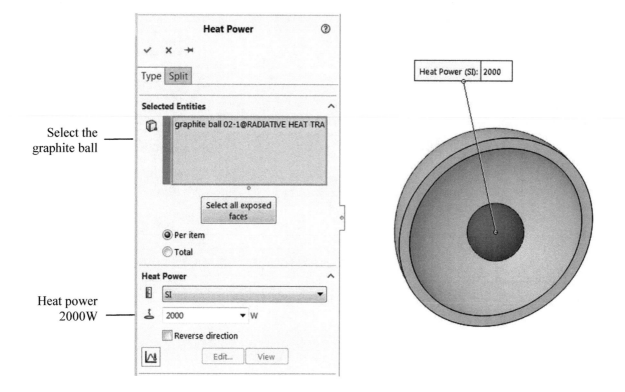

Select the
graphite ball

Heat power
2000W

Figure 16-17: Heat power definition.

Select assembly component graphite ball from the fly-out menu.

Now we assign a **Temperature** to the Reflector (Figure 16-18); this is done only to match the analysis setup in Flow Simulation where the default temperature of 293.2K was defined for all solids in Step 6 of the **Wizard** (Figure 16-6). This prescribed temperature has no impact on the temperature of the ball. The reflector serves only to reflect heat, and due to **Whitebody** surface conditions, does not absorb any heat.

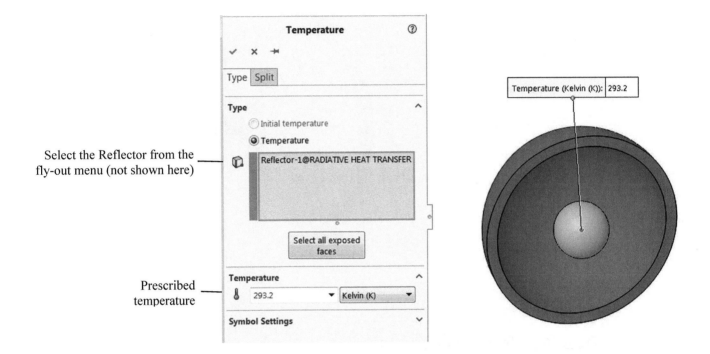

Figure 16-18: Temperature definition.

The temperature pointer appears to point to the graphite ball but it points to the origin of the coordinate system of the reflector.

Mesh the assembly using Curvature Based mesh, default element and run the solution. Create a **Temperature** plot and to make it look like the **Flow Simulation** plot, use **Section** plot as shown in Figure 16-18 and a temperature range from 1200K to the maximum.

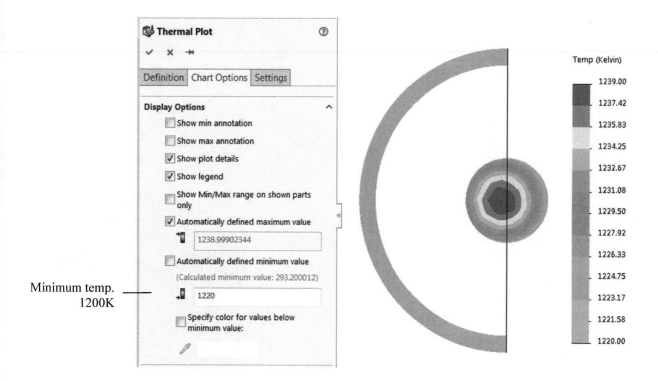

Figure 16-19: Temperature results plot.

This plot uses Section Clipping; Front reference plane serves as a section plane. Plot is shown in the section plane only.

Temperature results from the **Thermal** study in Figure 16-18 closely match the **Flow Simulation** results shown in Figure 16-14.

Summary of studies completed

Model	Configuration	Study or Project Name	Study Type
RADIATIVE HEAT TRANSFER.sldasm	*Default*	*FVM 01*	Flow Simulation
		FEM	Thermal

Figure 16-20: Names and types of studies completed in this chapter.

Notes:

17: NAFEMS Benchmarks

Topics covered

- ❏ Importance of benchmarks
- ❏ One dimensional heat transfer with radiation
- ❏ One dimensional transient heat transfer
- ❏ Two dimensional heat transfer with convection

Project description

Benchmarks are standard tests, which can be applied to any Finite Element system. They are intended for users who are interested in the performance of specific codes. The classic set of benchmarks was first published in 1986 by the National Agency for Finite Element Methods and Standards (NAFEMS – see chapter 20). NAFEMS benchmarks were designed to be a little more "searching" than those typically published by software developers since these naturally intended to demonstrate the favorable performance of a system rather than deficiencies.

In this chapter, we will test the performance of **SOLIDWORKS Simulation** in three NAFEMS thermal benchmarks with known analytical solutions. Performance will be evaluated by comparing **SOLIDWORKS Simulation** results with targets (expected results) as specified in each test.

Very simple geometries used in these tests lend themselves to more than one representation. Therefore, these exercises will also serve as a review of finite element modeling techniques available in thermal analysis with **SOLIDWORKS Simulation**.

Procedure

The thermal T2 tests code performance in steady state thermal analysis. A 1D bar has a constant prescribed temperature of 1000K applied to end A, and radiation to an ambient temperature of 300K to end B. There are no other boundary conditions present, meaning that the bar is insulated and there is no heat generated anywhere; the cross section is uniform (Figure 17-1).

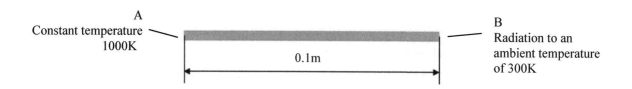

Figure 17-1: Geometry and boundary conditions of test T2.

Since this is a 1D model, only length needs to be specified.

Benchmark test T2 specifies material properties as follows:

> Conductivity = 55.6W/m/K
>
> Specific heat = 460J/kg/K
>
> Density = 7850kg/m³
>
> Emissivity at boundary B = 0.98

Notice that material specific heat and density are not required for a steady state thermal analysis.

The target in test T2 is temperature at end B. To pass the test, an FEA program should find this temperature to be equal to 927K.

This very simple geometry can be modeled in a number of ways, either as 3D or as a 2D problem. 1D elements are not available in **SOLIDWORKS Simulation**. Therefore, our choice is between 3D and 2D representations.

Should we decide to model this as a 3D problem, any prismatic bar with a uniform cross section along the length could be used (Figure 17-2).

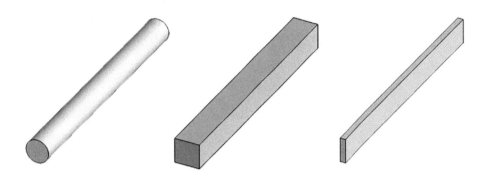

Figure 17-2: Examples of CAD geometry that could be used in a T2 test if 3D models were desired.

The shape and the area of the cross section are irrelevant in this 1D problem. Any of the above bars can be used as long as it has a length of 0.1m.

Should we decide to model this as a 2D problem, there is again more than one way to accomplish it. First, a bar with a round cross section could be used to create a 2D axisymmetric model. Second, a bar with a rectangular cross section could be used to create a 2D extruded model (Figure 17-3). Alternatively, a surface geometry could be prepared in **SOLIDWORKS** and treated as an axisymmetric or extruded model in a 2D study.

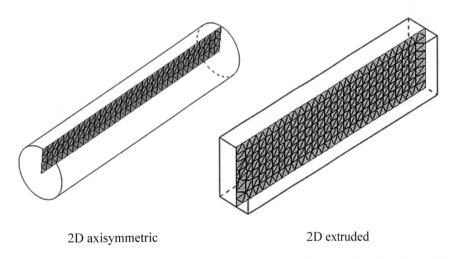

2D axisymmetric   2D extruded

Figure 17-3: Examples of CAD geometry that could be used in the T2 test if 2D models were desired.

Large finite elements are shown for clarity of this illustration. We will use much smaller elements for analysis.

For the T2 test, we will use a round bar and treat the problem as 3D. In test T3 which uses identical 1D geometry, we will use a square bar treating this problem as 2D extruded.

Open part model NAFEMS TEST T2 and verify that custom material required in test T2 has been defined. Create a steady state thermal study and apply thermal boundary conditions as shown in Figure 17-4 and Figure 17-5.

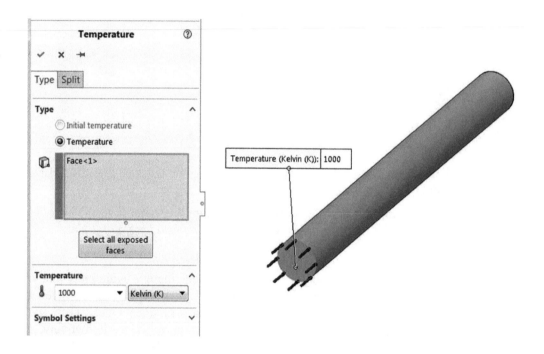

Figure 17-4: The boundary condition on side A is a prescribed temperature of 1000K.

A prescribed temperature is applied to side A. Prescribed temperature symbols are shown.

Create a custom material with properties specified in benchmark test T2 and apply it to the model. To define a custom material, refer to Figure 7-11.

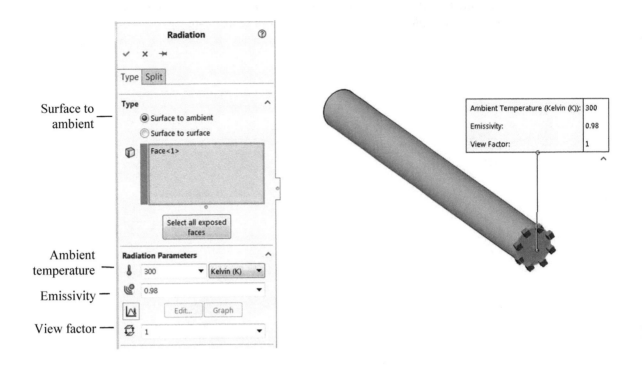

Surface to ambient

Ambient temperature

Emissivity

View factor

Figure 17-5: The boundary condition on side B is a Radiation Surface to ambient.

A radiation boundary condition is applied to side B. Enlarged radiation symbols are shown.

An **Ambient temperature** of 300K is the temperature of a distant enclosure, which receives heat radiated out from side B. An **Emissivity** of 0.98 is a material property. This is very close to the emissivity of black body so the material is very good at radiating out heat. To understand the meaning of a **View factor** of 1, take into consideration that **Surface to ambient** radiation is specified and that there is only one face in the model. A **View factor** equal to 1 means that the heat radiated out does not reach the face again.

Mesh the model with a default mesh and obtain a solution. Show the
Temperature plot and probe it as shown in Figure 17-6.

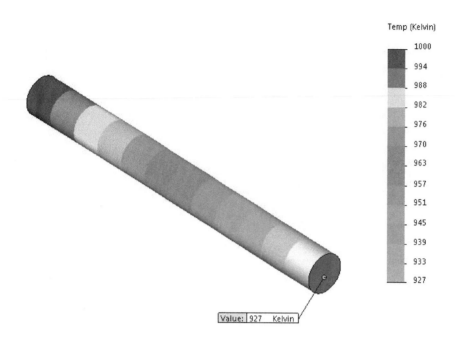

Figure 17-6: Probing shows the temperature on the side where the radiation
boundary condition has been defined.

The probed temperature is 927K.

Probing shows a temperature of 927K as specified in the benchmark target. This
demonstrates the correct performance of **Simulation** in this radiation problem.
Try running the same problem with a view factor of less than 1, which makes it
more difficult to radiate the heat out. The temperature will then be higher than
927K.

Benchmark test T3 requires a transient thermal analysis.

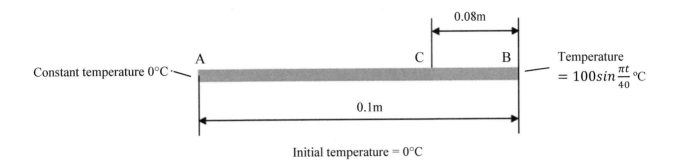

Figure 17-7: Geometry and boundary conditions of test T3.

The initial temperature is 0°C.

Benchmark test T3 specifies material properties as follows:

Conductivity = 35.0W/m/K

Specific heat = 440.5J/kg/K

Density = 7200kg/m^3

Model NAFEMS TEST T3.SLDPRT comes with these material properties defined.

Temperature on side B is a function of time; therefore, this is a transient thermal problem; specific heat and density are the required material properties.

The target in test T3 is a temperature at point C at time t=32s. To pass this benchmark test, an FEA program should find this temperature to be 36.6°C.

Open part model NAFEMS TEST T3 and define a thermal study using 2D simplification (Figure 17-8).

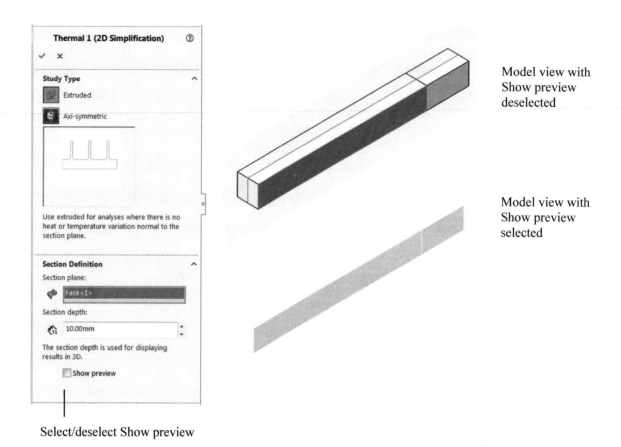

Model view with Show preview deselected

Model view with Show preview selected

Select/deselect Show preview

Figure 17-8: The 2D Simplification uses an Extruded 2D representation; thickness used here is 10mm but any thickness will satisfy benchmark test requirements.

Model is shown in two views; switch between these views by selecting/deselecting Show preview. Split lines marks location where temperature results will be probed.

Define study properties as shown in Figure 17-9.

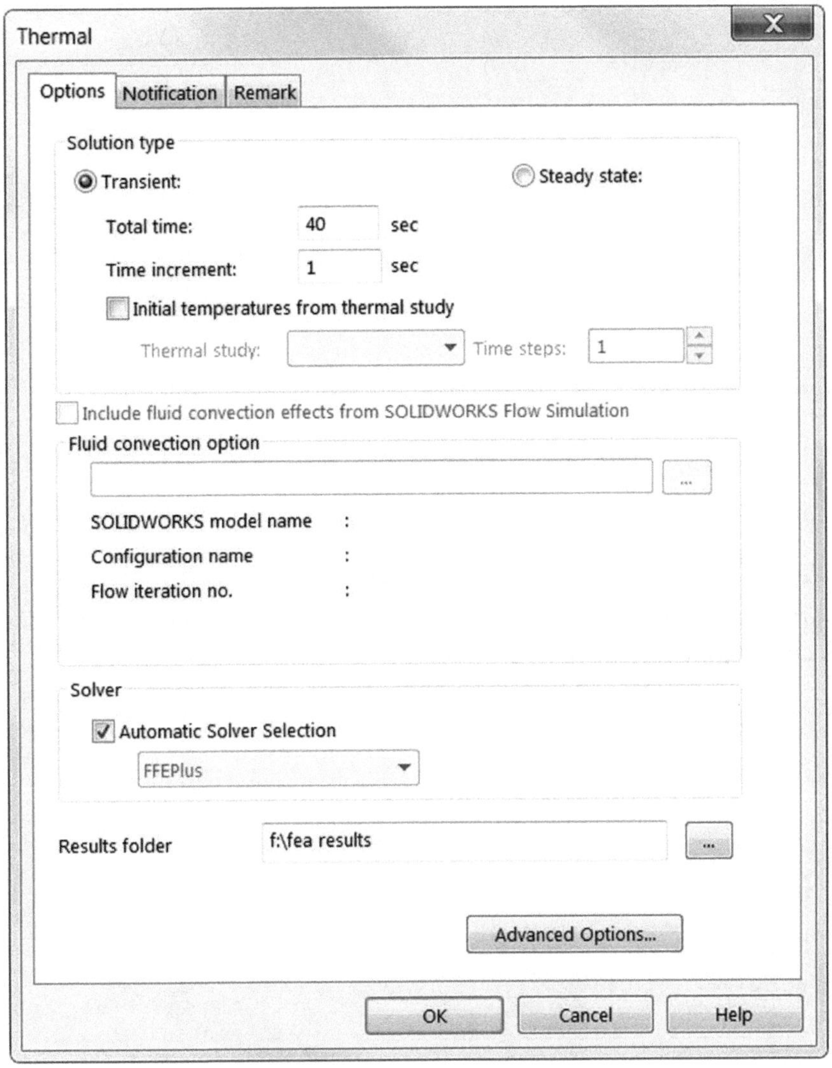

Figure 17-9: Properties of the transient thermal study.

The transient thermal study will be conducted in 40 steps with time increment of 1s.

Apply an **Initial Temperature** to faces as shown in Figure 17-10 and a **Prescribed Temperature** to side A as shown in Figure 17-11.

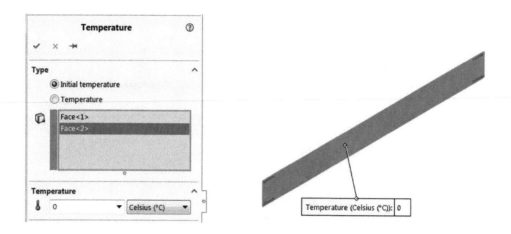

Figure 17-10: Definition of an initial temperature.

Initial temperature is applied to two faces; there are two faces in the 2D model because of the split line.

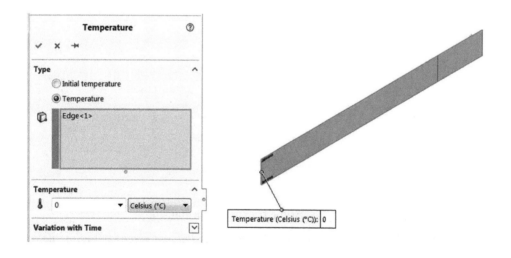

Figure 17-11: Definition of a prescribed temperature on side A.

Temperature is applied to the edge that represents a face in the 2D model.

Follow the steps in Figure 17-12 to define a time dependent temperature on side B.

(1)
Select edge on side B

(2)
Enter 1 as a multiplier of values in the Time curve table

(3)
Click Edit to open Time curve window

(4)
Copy the table from file NAFEMS TEST T3.xlsx and paste it here, click OK.

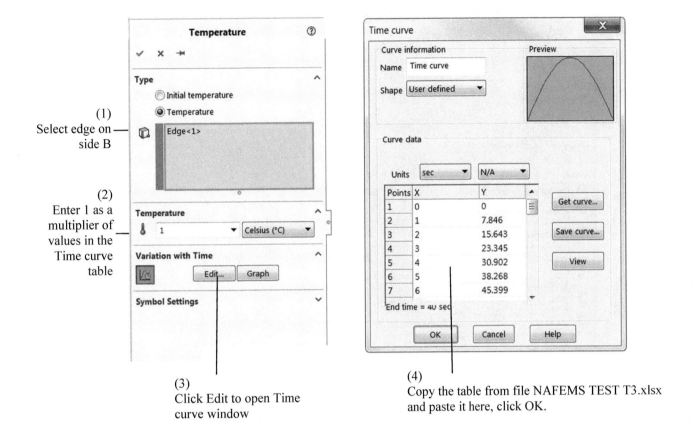

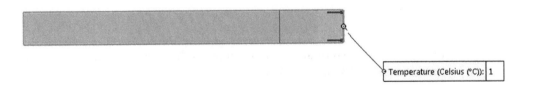

Figure 17-12: Definition of the time dependent temperature on side B.

The time dependency is illustrated by a curve shown in the Preview image.

The table in file T3.xlsx has been created by tabulating the equation shown in Figure 17-7.

Create and apply a new material to the geometry (conductivity = 35.0W/m/K, specific heat = 440.5J/kg/K, density = 7200kg/m^3). Mesh the model with the default element size and obtain a solution. To check the target, probe **Temperature** plot as shown in Figure 17-13.

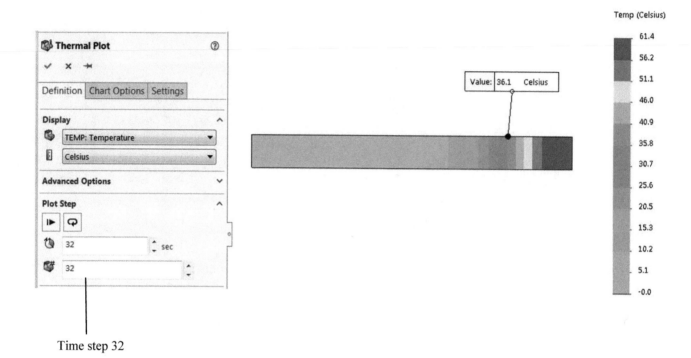

Time step 32

Figure 17-13: Probing temperature plot in location C.

Make sure the plot is for time 32s and probing is where split line separates the two faces.

Probing shows a **Simulation** temperature result of 36.1°C, very close to the benchmark test T3 target of 36.6°C.

Benchmark test T4 focuses on conductive heat transfer induced by prescribed temperatures and convection to ambient fluid (Figure 17-14).

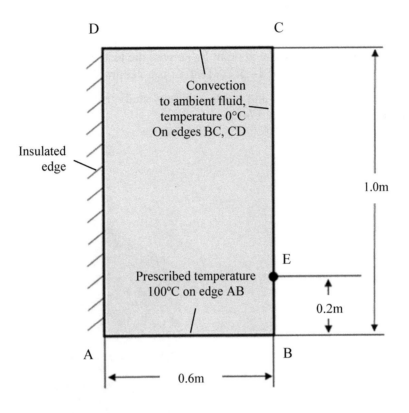

<u>Figure 17-14: Geometry and boundary conditions of test T4.</u>

This is a 2D problem; the thickness is uniform.

Benchmark test T4 specifies a material conductivity and convection properties as follows:

Conductivity = 52.0W/m/K

Convection = 750W/m^2/K

The target in test T4 is temperature at point E. To pass the test, an FEA program should find this temperature to be 18.3°C. Test T4 is a 2D problem.

Open model NAFEMS TEST T4 and create a thermal study using 2D Simplification as shown in Figure 17-15.

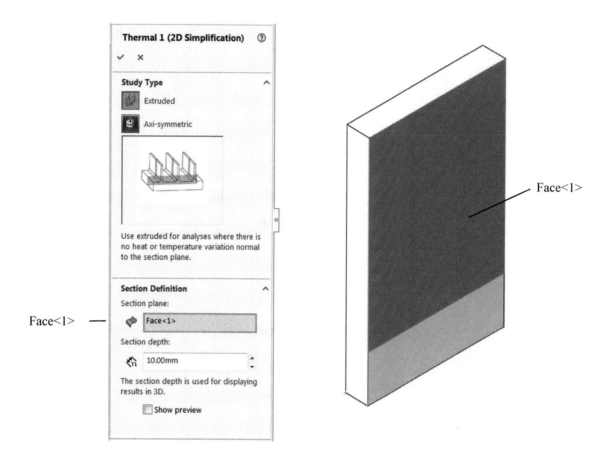

Figure 17-15: 2D Simplification uses Extruded 2D representation.

This is a 2D extruded problem; the thickness is uniform. Any thickness will satisfy benchmark test requirements.

Apply prescribed temperature and convection boundary conditions as shown in Figure 17-16 and Figure 17-17.

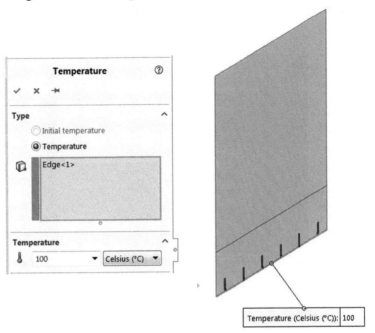

Figure 17-16: Definition of prescribed temperature in test T4.

Apply a temperature of 100°C to the lower edge. Prescribed temperature symbols are shown.

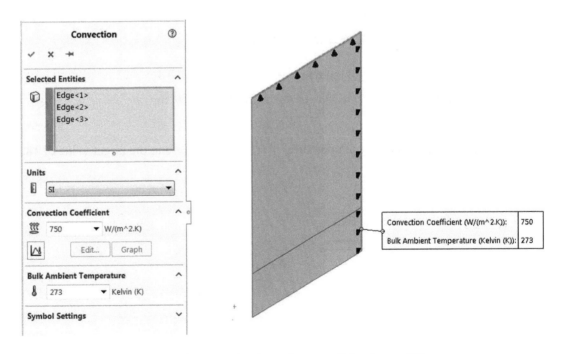

Figure 17-17: Definition of convection boundary condition in test T4.

Apply a convection of 750W/m²/K and an ambient temperature of 273K to the top and right edges. Convection symbols are shown.

Verify that the custom material defined in *T4* model has thermal conductivity equal to 52.0W/m/K. Mesh the model with a default mesh and obtain a solution. Probe the **Temperature** plot at point E as shown in Figure 17-18.

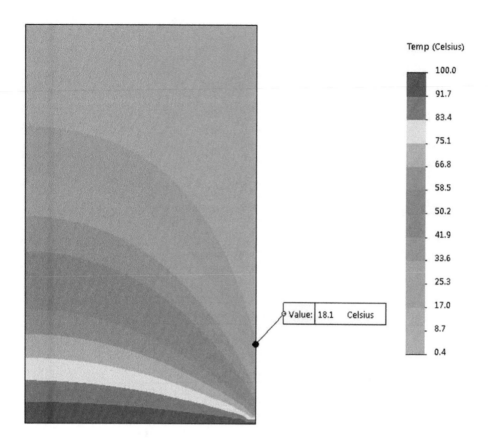

Figure 17-18: Probing temperature results in test T4.

Probe at the intersection of the vertical edge with the horizontal split line.

Probing shows a temperature of 18.1°C, very close to the target value of 18.3°C.

Summary of studies completed

Model	Configuration	Study Name	Study Type
NAFEMS TEST T2.sldprt	*Default*	*Thermal 1*	Thermal
NAFEMS TEST T3.sldprt	*Default*	*Thermal 1*	Thermal
NAFEMS TEST T4.sldprt	*Default*	*Thermal 1*	Thermal

Figure 17-19: Names and types of studies completed in this chapter.

Notes:

18: Summary and miscellaneous topics

Topics covered

- ❑ Summary of thermal analysis exercises
- ❑ Nonlinear transient problems
- ❑ Advanced options of thermal study
- ❑ Closing remarks

As shown in the flowchart in Figure 18-1, thermal analysis may be linear or nonlinear, steady state or time dependent.

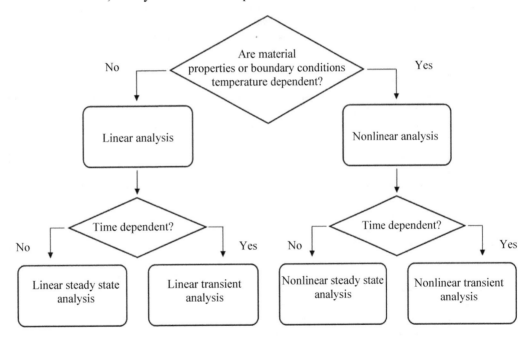

Figure 18-1: Classification of thermal analysis with regard to time dependency and nonlinearity.

This is a repetition of Figure 1-12.

In linear analyses, the conductivity matrix does not change during the solution process; in nonlinear analyses it must be modified. In steady state analyses, the boundary conditions do not change during the solution process; in time dependent (transient) analyses, boundary conditions change.

The table in Figure 18-2 classifies examples discussed in chapters 1-13 in terms of nonlinearity and time dependency.

Chapter	Model	Linear steady state analysis	Linear transient analysis	Nonlinear steady state analysis	Nonlinear transient analysis
1	BRACKET TH	yes			
1	HEAT SINK 01	yes			
1	DOUBLE SYM	yes			
1	AXI SYM	yes			
2	HOLLOW PLATE TH	yes			
3	L BRACKET TH	yes			
4	STEEL ROD	yes			
4	TWO RODS	yes			
5	HEATING DUCTS	yes			
6	HEATING DUCTS	yes			
7	HOT PLATE	yes	yes		
8	MUG	yes	yes		
9	RECT BAR	yes			
10	HEAT SINK	yes	yes		
11	GRAPHITE BALL			yes	
12	HEMISPHERE			yes	
13	BLOCKS			yes	
16	NAFEMS TEST T2			yes	
16	NAFEMS TEST T3		yes		
16	NAFEMS TEST T4	yes			

Figure 18-2: Classification of exercises according to time dependency and nonlinearity.

When "yes" appears both in steady state and transient analysis, it means that more than one study was conducted.

All radiation problems are nonlinear because a radiation boundary condition is a function of temperature in fourth power. However, none of the exercises discussed so far presented a transient and nonlinear problem.

What would make a transient, nonlinear problem? This could be any radiation problem with at least one time dependent boundary condition. This also could be any transient problem with, for example, a temperature dependent material property. We will analyze both problems starting with a radiation problem which can be seen as a continuation of radiation topics presented in chapters 11, 12, 13 and 16.

Open assembly model SPACE HEATER. The model consists of an aluminum Reflector and a Glass bulb. Faces that exchange heat by radiation are indicated in Figure 18-3, Front view. The rim and back face are assumed to be insulated. A fillet at the base of the stem is also insulated; no boundary conditions are defined there. This is done to simplify calculations of view factors and this omission won't have much of an effect because of the small surface area of the fillet.

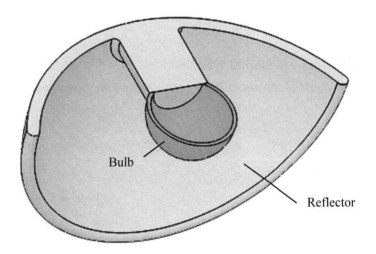

Figure 18-3: SPACE HEATER assembly model has two components: Reflector and Bulb.

The model is shown in a section view to demonstrate that Bulb is hollow.

Radiation is the only mechanism of removing heat from this model. Conduction is present but heat traveling by conduction through the Bulb and Reflector is also removed by radiation. Create a **Steady State** thermal study called *01 steady state*. Define **Heat Power** and **Radiation** conditions as shown in Figures 18-4, 18-5, and 18-6.

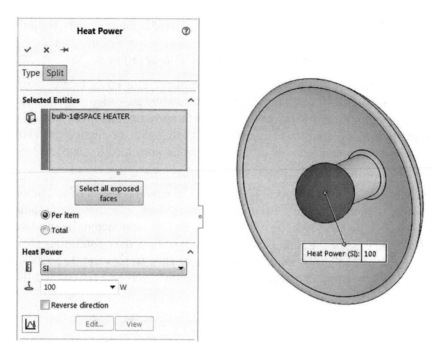

Figure 18-4: Heat power definition.

Apply 100W to the entire Bulb; select it from the fly-out menu not shown here.

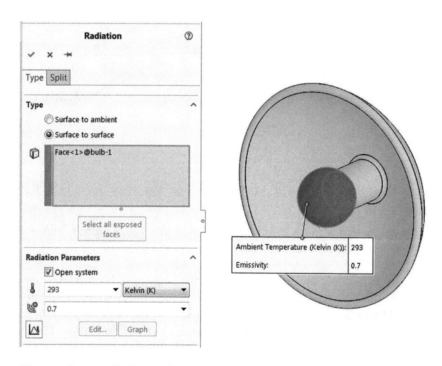

Figure 18-5: Radiation definition on the Bulb surface.

Open system means that some heat is radiated out to space and not to the Reflector. The ambient temperature is 293K, and emissivity of Bulb surface material is 0.7.

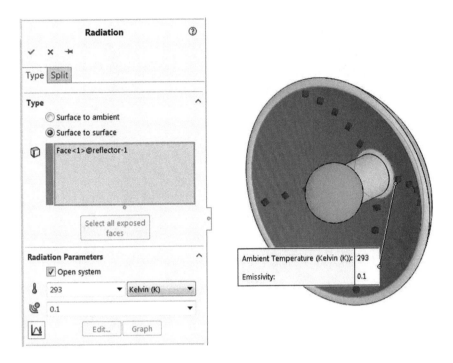

Figure 18-6: Radiation definition on Reflector surface and stem.

Open system means that some heat is radiated out to space and not all to the Bulb. Ambient temperature is 293K, and emissivity of these faces is 0.1. Note that the Reflector is designed to reflect radiation, which is why the value of the emissivity is low.

Enlarged radiation symbols are shown.

Mesh the assembly with the default element size, which produces a mesh barely acceptable for temperature analysis. Analysis of heat flux would definitely need a more refined mesh especially at the area of the fillet.

Run the solution and review the steady state temperature as shown in Figure 18-7.

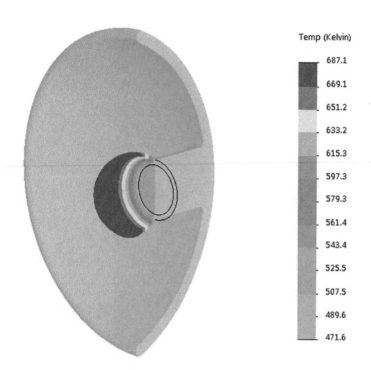

Figure 18-7: Steady state temperature results.

A section plot is used in this illustration.

Even though the geometry has axial symmetry, the radiation problem can't be simplified to a 2D axisymmetric model. This is because finding **View Factors** requires 3D geometry.

Notice that we have just solved nonlinear, steady state problem. We will now make it into a transient problem. Copy study *01 steady state* into *02 transient*. Define properties of the transient study as shown in Figure 18-8.

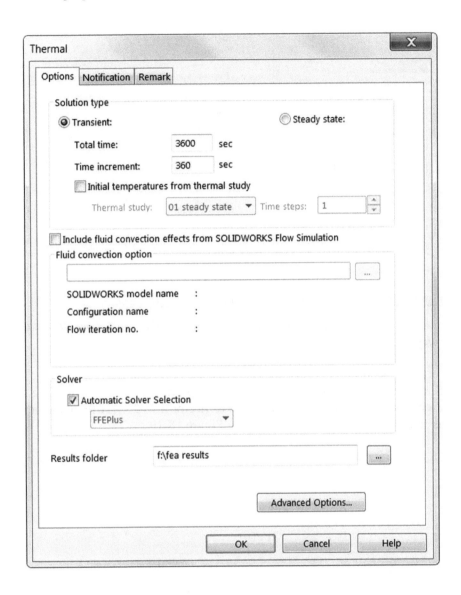

Figure 18-8: Properties of the transient thermal study.

The total time is 1 hour covered in 10 steps at 6min intervals.

Remember that transient thermal analysis requires an **Initial Temperature**. Define it as 298K for the entire assembly (Figure 18-9).

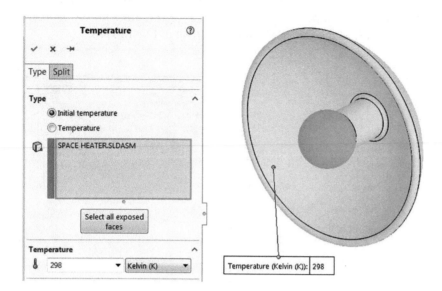

Figure 18-9: Definition of Initial Temperature for the assembly.

Select the assembly from the fly-out menu. Notice that the temperature callout refers to the origin of the assembly.

Run study *02 transient* and display the **Temperature** plot from any time step. Probe the temperature of the Bulb and Reflector then display these temperatures as function of time; follow the steps in Figure 18-10.

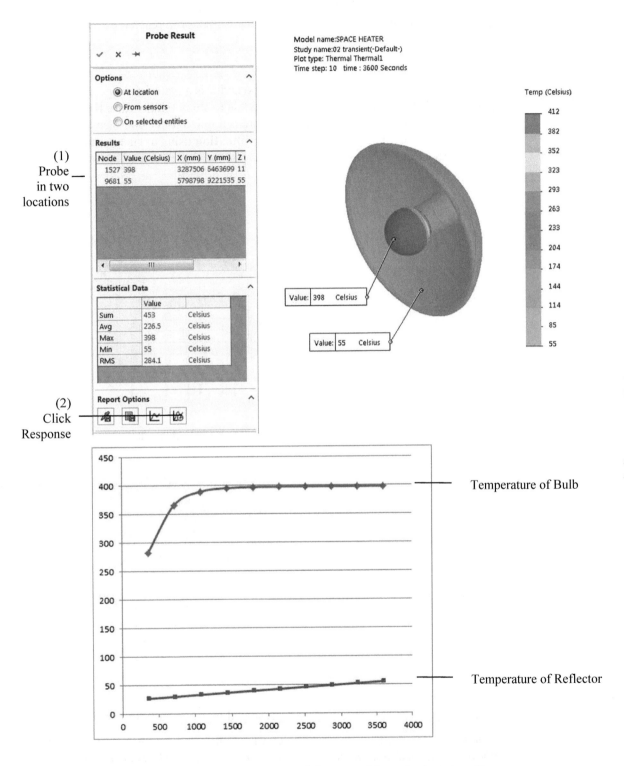

Figure 18-10: Plot shows temperature of the assembly at the last step. Response graph shows temperatures in the probed location as function of time.

The temperature of the front side of the Bulb and the temperature of the Reflector are almost uniform; you may probe just about anywhere.

Results indicate that the Bulb reaches steady state temperature after about 1500s. The duration of analysis was too short for the Reflector to reach the steady state temperature.

In the next model the source of nonlinear behavior is a material property. Open part model RADIATOR AXISYM. Our goal is to study temperature fluctuation on the ribs when a time variable temperature is applied to the end face of the stem. Create a thermal study with **2D Simplification** using axial symmetry (Figure 18-11).

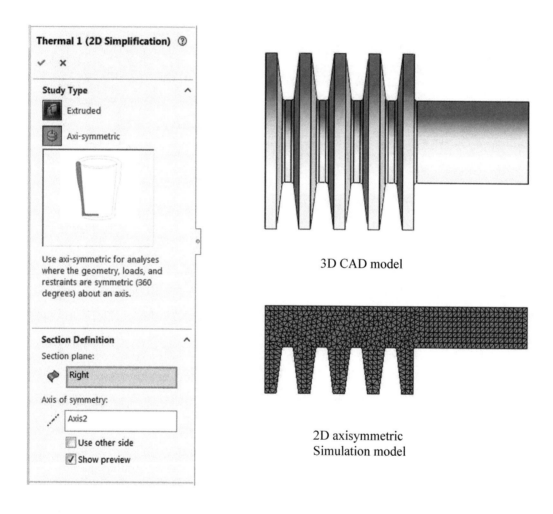

3D CAD model

2D axisymmetric
Simulation model

<u>Figure 18-11: The 3D model can be represented in 2D as half of the radial cross.</u>

Using 2D elements, we can use a fine mesh with no need for mesh control. Shown above is a mesh that used the default element size. The mesh is acceptable for temperature analysis but may require refinement for heat flux analysis.

2D Simplification window shows Section plane and Axis of symmetry used to create the 2D model.

To create material with temperature dependent conductivity, first copy the material 1060 Alloy into **Custom Materials** in the **Thermal** folder, then follow the steps in Figure 18-12.

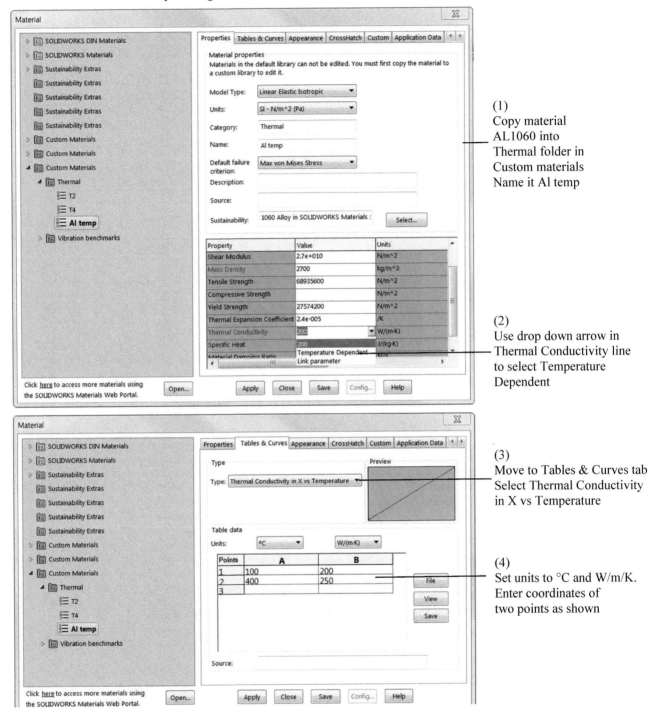

(1)
Copy material AL1060 into Thermal folder in Custom materials Name it Al temp

(2)
Use drop down arrow in Thermal Conductivity line to select Temperature Dependent

(3)
Move to Tables & Curves tab Select Thermal Conductivity in X vs Temperature

(4)
Set units to °C and W/m/K. Enter coordinates of two points as shown

Figure 18-12: Definition of custom material with temperature dependent conductivity.

Define study properties as shown in Figure 18-13.

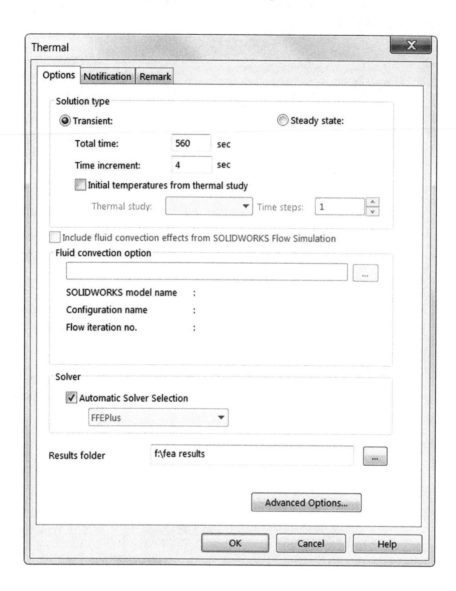

<u>Figure 18-13: Definition of properties of the transient study.</u>

The total duration is 560s; the time increment is 4s. The analysis will be conducted in 140 steps.

Define an Initial Temperature of 0°C as shown in Figure 18-14.

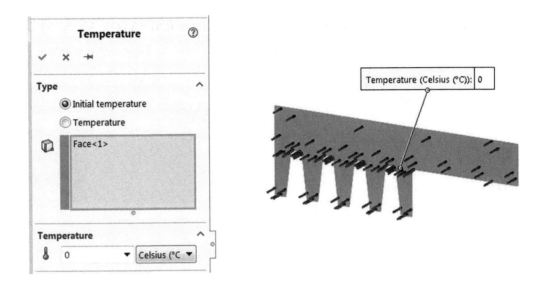

Figure 18-14: An initial temperature of 0°C is applied to the face that represents model in the 2D simplification.

An oblique view is used to display temperature symbols.

Define the time dependent temperature to the right end. The time dependency follows the equation:

$$t = 100 sin \frac{\pi t}{20} + 100 \text{ °C}.$$

This equation has been tabulated in the spreadsheet RADIATOR AXISYM.xlsx.

You may copy the table from there and paste it into the Time curve window.

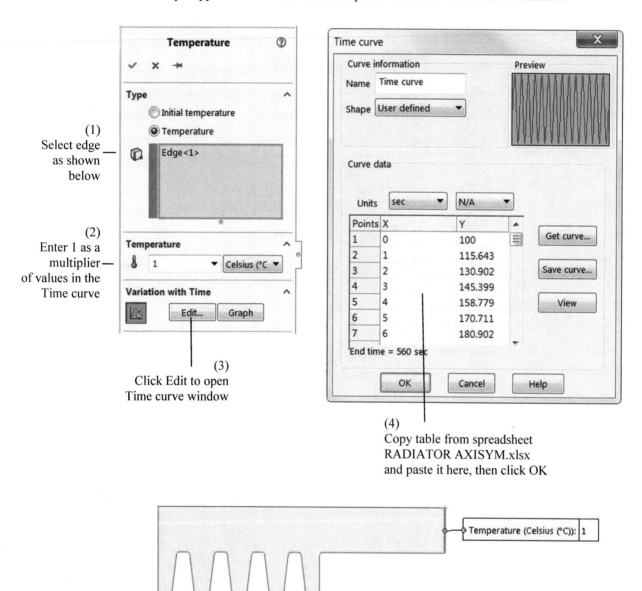

(1) Select edge as shown below

(2) Enter 1 as a multiplier of values in the Time curve

(3) Click Edit to open Time curve window

(4) Copy table from spreadsheet RADIATOR AXISYM.xlsx and paste it here, then click OK

Figure 18-15: Definition of time dependent temperature.

These steps are similar to those in Figure 17-12.

Mesh the model with the default mesh (Figure 18-11) and obtain solution. Display the **Temperature** plot from any step and probe results in two locations as shown in Figure 18-16.

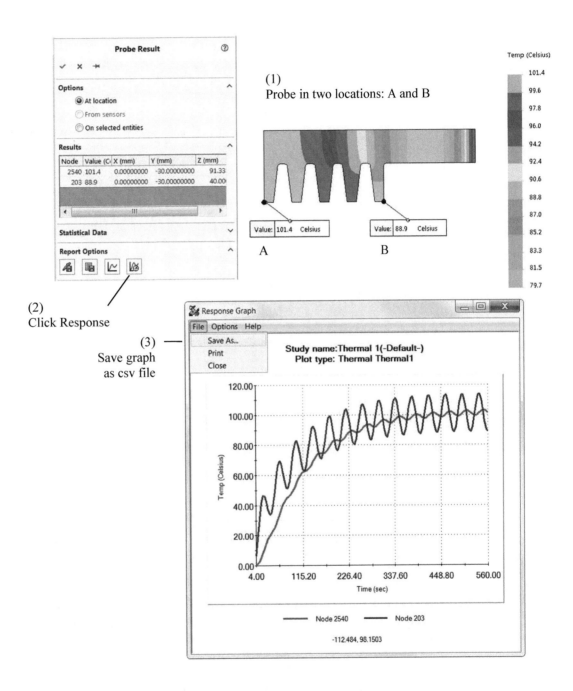

Figure 18-16: Probe temperature results in two locations as shown.

Save the graph as a csv file.

Graph results are easier to present using Excel, which offers better graphing tools than those in **SOLIDWORKS Simulation**. The prescribed temperature and temperatures in the two probed locations is shown as a function of time in Figure 18-17.

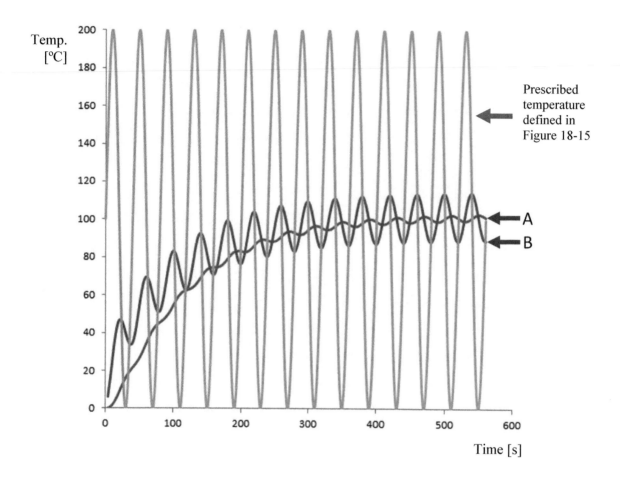

Figure 18-17: Probed temperature results in locations A and B.

This graph also shows the prescribed temperature oscillating between 0° and 200°C.

As we can see in Figure 18-17, temperature oscillations diminish as we move further away from where the oscillating prescribed temperature is applied. This is the effect of thermal inertia.

Notice that the RADIATOR AXISYM problem is transient and nonlinear. It solves fast thanks to the 2D representation that greatly reduces effort of numerical solution. Benefits of the 2D representation are particularly noticeable when transient and/or nonlinear problems are solved.

Advanced options of Thermal study

Thermal study offers **Advanced Options** that control solution of nonlinear problems (Figure 18-18).

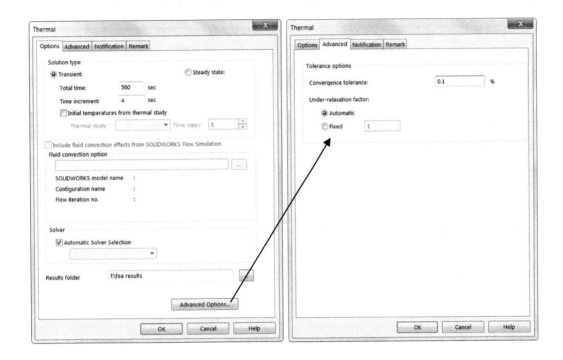

Figure 18-18: Advanced Option in Thermal study.

Use Advanced Options to control convergence of nonlinear problem solution.

These settings are used by thermal problems solver in a similar manner to how **Advanced Options** are used in nonlinear static studies.

The advanced thermal options are integrated in both the Direct sparse and FFEPlus solvers. For example, the Under-Relaxation Factor in Thermal **Advanced Options** is analogous to the **Singularity Elimination Factor** in **Advanced Options** of **Nonlinear** study.

Advanced Options are particularly useful in radiation problems.

Summary of studies completed

Model	Configuration	Study Name	Study Type
SPACE HEATER.sldasm	*Default*	*01 steady state*	Thermal
		02 transient	Thermal
RADIATOR AXISYM.sldprt	*Default*	*Thermal 1*	Thermal

Figure 18-19: Names and types of studies completed in this chapter.

19: Glossary of terms

The following glossary provides short descriptions of selected terms used in this book.

****

Heat
Heat is a form of energy of a system when a temperature difference exists between a system and its surroundings.

Heat transfer
Heat transfer is the transfer of heat energy from one system to another.

Conjugate heat transfer
Heat transfer within solids and fluids, due to thermal interaction between the solids and fluids.

Temperature
Temperature is a measure of heat energy. It can be measured in K (Kelvin), °C(Celsius), and °F(Fahrenheit). Notice that the degree symbol ° is not used with K.

Heat flux
Heat flux is the rate of heat transfer per unit area.

Conduction
Conduction is a mode of heat transfer in which the heat energy is transferred on a molecular scale with no movement of macroscopic particles. Fourier's Law characterizes heat flow by conduction. The coefficient of thermal conduction characterizes heat transfer in solid bodies.

Convection
Convection is a mode of heat transfer between a solid face and a fluid. The movement of macroscopic particles transfers the heat energy. The convection coefficient characterizes convection between a solid body and its surrounding fluid.

Radiation
Radiation heat transfer occurs by means of electromagnetic waves. The heat energy can be transferred in vacuum. The Stefan-Boltzmann law characterizes radiative heat transfer.

Linear analysis
An analysis is linear when none of the parameters characterizing boundary conditions, material properties and loads is temperature dependent.

Nonlinear analysis
An analysis is nonlinear when at least one of the parameters characterizing boundary conditions, material properties or loads is temperature dependent.

Steady state analysis
In steady state analysis none of the parameters characterizing boundary conditions, material properties or loads is time dependent.

Transient analysis
An analysis is transient if at least one of the parameters characterizing boundary conditions, material properties or load is time dependent.

Conductivity
Conductivity characterizes heat flow per unit area in the normal direction in a solid body which takes place as a result of a temperature difference.

Specific heat
Specific heat is the energy required to raise the temperature of a unit of mass through a one degree Celsius (or Kelvin) temperature rise.

Emissivity
Emissivity is the ratio of the total energy emitted by a surface to the total energy emitted by a black surface at the same temperature.

View factor
View factor is a ratio of the energy reaching the second surface to the energy emitted by the first surface. If only one surface is involved and it is radiating all of its heat out, the view factor accounts for the fact that some radiation hits the same surface again before being finally radiated out to space.

Thermal stress
Thermal stress is stress induced in a structure as a result of non-uniform temperature distribution in the structure.

Thermal expansion
Thermal expansion is the tendency of matter to change in a volume in response to a change in temperature. It is described by a coefficient of thermal expansion.

20: References

This textbook builds on an understanding of structural analysis with the Finite Element Method and assumes familiarity with structural and introductory thermal analysis to the extent of the textbook (1).

To review fundamentals of Heat Transfer and fundamentals of the Finite Element Method, readers are referred to the following literature listed here in the order of relevance.

1. Kurowski P. "Engineering Analysis with SOLIDWORKS Simulation 2017", SDC Publications

2. NAFEMS Background to Benchmarks, NAFEMS 1993

3. How to undertake Finite Element Based Thermal Analysis NAFEMS 2002

4. Incropera F., Dewitt D., Bergman T., Lavine A. "Fundamentals of Heat and Mass Transfer", John Wiley & Sons, Inc.

5. Logan D. "A First Course in the Finite Element Method", Brooks/Cole

The full list of NAFEMS publications can be found at www.nafems.org. Another internet site with a number of FEA related publications is presented by Design Generator Inc. Publications related to FEA fundamentals, training, and implementation can be found at www.designgenerator.com/publications.htm.

21: List of exercises

Chapter	Part	Assembly	Spreadsheet
1	BRACKET TH HEAT SINK 01 DOUBLE SYM AXISYM		
2	HOLLOW PLATE TH		
3	L BRACKET TH		
4	STEEL ROD	TWO RODS	
5		HEATING DUCTS	
6		HEATING DUCTS	
7		HOT PLATE	
8	MUG		
9	LINK		
10		HEAT SINK TH	HEAT SINK TH.xlsx
11	GRAPHITE BALL		
12	HEMISPHERE		
13		BLOCKS	
14		ELBOW HEATER	
15		INTERNAL FLOW* EXTERNAL FLOW* BALL15	
16		RADIATIVE HEAT TRANSFER	
17	NAFEMS TEST T2 NAFEMS TEST T3 NAFEMS TEST T4		NAFEMS TEST T3.xlsx
18	RADIATOR AXISYM	SPACE HEATER	RADIATOR AXISYM.xlsx

Figure 21-1: SOLIDWORKS models and Excel spreadsheets that accompany this book.

* Model comes with Flow Simulation project defined.

Notes: